Jürgen Pfingstner

Mitigation of ground motion effects via feedback systems in CLIC

Jürgen Pfingstner

Mitigation of ground motion effects via feedback systems in CLIC

Südwestdeutscher Verlag für Hochschulschriften

Impressum / Imprint

Bibliografische Information der Deutschen Nationalbibliothek: Die Deutsche Nationalbibliothek verzeichnet diese Publikation in der Deutschen Nationalbibliografie; detaillierte bibliografische Daten sind im Internet über http://dnb.d-nb.de abrufbar.

Alle in diesem Buch genannten Marken und Produktnamen unterliegen warenzeichen-, marken- oder patentrechtlichem Schutz bzw. sind Warenzeichen oder eingetragene Warenzeichen der jeweiligen Inhaber. Die Wiedergabe von Marken, Produktnamen, Gebrauchsnamen, Handelsnamen, Warenbezeichnungen u.s.w. in diesem Werk berechtigt auch ohne besondere Kennzeichnung nicht zu der Annahme, dass solche Namen im Sinne der Warenzeichen- und Markenschutzgesetzgebung als frei zu betrachten wären und daher von jedermann benutzt werden dürften.

Bibliographic information published by the Deutsche Nationalbibliothek: The Deutsche Nationalbibliothek lists this publication in the Deutsche Nationalbibliografie; detailed bibliographic data are available in the Internet at http://dnb.d-nb.de.

Any brand names and product names mentioned in this book are subject to trademark, brand or patent protection and are trademarks or registered trademarks of their respective holders. The use of brand names, product names, common names, trade names, product descriptions etc. even without a particular marking in this works is in no way to be construed to mean that such names may be regarded as unrestricted in respect of trademark and brand protection legislation and could thus be used by anyone.

Coverbild / Cover image: www.ingimage.com

Verlag / Publisher:
Südwestdeutscher Verlag für Hochschulschriften
ist ein Imprint der / is a trademark of
AV Akademikerverlag GmbH & Co. KG
Heinrich-Böcking-Str. 6-8, 66121 Saarbrücken, Deutschland / Germany
Email: info@svh-verlag.de

Herstellung: siehe letzte Seite /
Printed at: see last page
ISBN: 978-3-8381-3692-9

Zugl. / Approved by: Wien, TU, Diss., 2013

Copyright © 2013 AV Akademikerverlag GmbH & Co. KG
Alle Rechte vorbehalten. / All rights reserved. Saarbrücken 2013

Acknowledgement

Before the outcome of the work of the last three years is presented, it is time to thank the people who directly or indirectly contributed to this work.

The first person I want to mention is at the same time the first person I met at CERN. I still remember the day of my first interview, standing insecurely in front of the reception building, seeing the tall man in leather trousers walking towards me. Hermann, thank you very much for making this thesis possible and for integrating me so warmly into the community at CERN.

My supervision was soon overtaken by Daniel Schulte. I would like to thank him for his very helpful support and also for the generous financial support for the work on this thesis. Even more important for me was our collaborative work in which I could look over the shoulder of a great scientist. It was amusing and enlightening at the same time, to see what football has to do with science and Greek philosophy with quantum mechanics.

I also want to thank my academic supervisor Michael Benedikt, for taking his time to help me even though his schedule was already crowded. His handling of the university activities was very efficient and professional.

Next, I want to thank my work colleagues for the many very interesting and fruitful discussions. Especially, I want to point out Jochem Snuverink with whom I formed a working team for almost two years. His influence on this work cannot be overstated. He helped with the implementation of the simulation framework, the simulations itself and via countless hours of discussions. He was also the poor guy who proofread this thesis first.

Since productive work is only possible if one can refill his energy buffers, I want to thank my battery chargers. I thank my coffee team, Claus, Fadmar, Matthias, Øystein, Stefano and Patrik for the extensive discussions about the European football scene, which I had no clue about, but still enjoyed a lot. Further, I want to thank Johanna, Jean-Christophe, Iulia, Ulrich, Stefan, Benjamin, Verena, Georg, Dan (thanks also for checking my English), Harald and all the other friends in Geneva for making my stay so enjoyable.

Before I focus my attention to people at home in Austria, I mention the two most important persons for me here in France/Switzerland. Arno, to thank you for something specific would be nothing but ridiculous considering that many friends say we behave like an old couple. Therefore, I just thank you collectively for being a good friend for 16 years. Cornelia, you gave me a very good reason to leave my office not too late in the evening and also made me get up in the morning with Swiss precision. But much more you helped me with the wonderful and unforgettable time I could spend with you.

Next, I want to thank my relatives and friends in my home village Perchau. At my visits, I was always amazed how they could give me the feeling of complete integration, as if I would have been only on vacation for maybe one week. I especially want to mention my three little cousins Kerstin, Carina and Birgit. The same is true for my friends in Graz, where I want to point out the hospitality of Daniel, Susi and Tomi during my stays.

A special remark I make for my aunt Monika and my uncle Manfred, for their never-ending care and their strong influence on my professional career. You have been a fruitful ground for my interests at an early age, by explaining me things that other people thought I was much too young for.

My biggest thanks go to my family. My brothers Gerd and Hannes, and also their girlfriends Alexandra and Teresa, I thank for many things, but with respect to this thesis especially for grounding me and preventing me from becoming a nerd. To my mother, I am grateful for her effort to keep close contact in spite of the far distance and even more for her beaming face whenever I enter our house in Perchau. Finally, I thank my father, who formed my working attitude by being an idol to me in diligence, endurance, stability but also open-mindedness throughout my life.

Jürgen Pfingstner, February 2013

Abstract

The Compact Linear Collider (CLIC) is a future multi-TeV electron positron collider, which is currently being designed at CERN. To achieve its ambitious goals, CLIC has to produce particle beams of the highest quality, which makes the accelerator very sensitive to ground motion. Four mitigation methods have been foreseen by the CLIC design group to cope with the feasibility issue of ground motion. This thesis is concerned with the design of one of these mitigation methods, named linac feedback (L-FB), but also with the simultaneous simulation and validation of all mitigation methods. Additionally, a technique to improve the quality of the indispensable system knowledge has been developed.

The L-FB suppresses beam oscillations along the accelerator. Its design is based on the decoupling of the overall accelerator system into independent channels. For each channel an individual compensator is found with the help of a semi- automatic control synthesis procedure. This technique allows the designer to incorporate expert knowledge, which is used by an optimisation algorithm to minimise the luminosity loss due to ground motion. This approach speeds up the design process significantly, while at the same time improving the orbit feedback performance compared to standard methods. Beside the L-FB, simple but effective designs for the interaction point feedback and cost reduction options for the quadrupole stabilisation are presented. For the design of all these feedback systems models of the ground motion influence on different beam parameters such as beam offset, beam size and luminosity have been derived by adapting and extending existent models.

To design, improve and validate the ground motion mitigation methods, a simulation framework was set up, which includes a ground motion generator, beam tracking, beam-beam interaction and all mitigation methods. The simulations show that the ground motion mitigation methods can efficiently preserve the CLIC luminosity. Due to our design of the L-FB, the specifications of the beam position monitor resolution could be relaxed significantly. The robustness of the L-FB was also verified with respect to many other imperfections. Only a certain sensitivity to beam energy variations was observed, which could be resolved by filtering dispersive orbits from the measurements. Further simulation results were an essential input for the redesign of the quadrupole stabilisation system leading to a significant performance improvement of the system.

Due to the high importance of the system knowledge for many applications, a system identification scheme was developed. It is capable of adapting the parameters of the system model to changes of the main linac behaviour (orbit response matrix) on-line, during the regular operation of the linac. By focusing only on the most significant system changes, the identification speed could be improved strongly compared to standard algorithms. The identified parameters of the orbit response matrix can be used to improve the performance of beam-based alignment algorithms and orbit feedbacks and are an important input for diagnosis and error detection tools.

List of publications

Primary publications

I	**Adaptive Scheme for the CLIC Orbit Feedback** J. Pfingstner, D. Schulte and M. Hofbaur In Proc. of the 1st Int. Particle Acc. Conf. (IPAC10), 2010
II	**Amplitude model for beam oscillations in the main linac of CLIC** J. Pfingstner, D. Schulte, H. Schmickler and M. Hofbaur CERN-OPEN-2011-010, CLIC-Note-860, CERN, 2010
III	**An interleaved, model-supported system identification scheme for the particle accelerator CLIC** J. Pfingstner, D. Schulte, H. Schmickler and M. Hofbaur In Proc. of the 49th IEEE Conf. on Decision and Control (CDC10), 2010
IV	**SVD-based filter design for the orbit feedback of CLIC** J. Pfingstner , M. Hofbaur, D. Schulte and J. Snuverink In Proc. of the 2nd Int. Particle Acc. Conf. (IPAC11), 2011
V	**Lockerung von Sensortoleranzen mittels regelungstechnischer Methoden für den Teilchenbeschleuniger CLIC** J. Pfingstner, J. Snuverink, D. Schulte and M. Hofbaur In 17. Steirisches Seminar über Regelungstechnik und Prozessautomatisierung, 2011
VI	**Recent improvements in the orbit feedback and ground motion mitigation techniques for CLIC** J. Pfingstner, J. Snuverink, A. Latina and D. Schulte In Proc. of the 3nd Int. Particle Acc. Conf. (IPAC12), 2012
VII	**Ground motion optimised orbit feedback design for the future linear collider** J. Pfingstner, J. Snuverink and D. Schulte Nucl. Instrum. and Meth. A, 703:163-170, 2013

Secondary publications

i   **Study of the stabilization to the nanometer level of mechanical vibrations of the CLIC main beam quadrupoles**
K. Artoos et al.
Proc. of the 2009 Particle Acc. Conf. (PAC09)

ii   **Sub-nm beam motion analysis using a standard BPM with high resolution electronics**
M. Gasior et al.
Proc. of the 2010 Beam Instr. Workshop (BIW10)

iii   **Status of Ground Motion Mitigation Techniques for CLIC**
J. Snuverink et al.
Proc. of the 2nd Int. Particle Acc. Conf. (IPAC11), 2011

iv   **System control for the CLIC main beam quadrupole stabilization and nano-positioning**
St. Janssens et al.
Proc. of the 2nd Int. Particle Acc. Conf. (IPAC11), 2011

v   **Interaction point feedback design and integrated simulations to stabilize the CLIC final focus**
G. Balik et al.
Proc. of the 2nd Int. Particle Acc. Conf. (IPAC11), 2011

vi   **BDS tuning and luminosity monitoring in CLIC**
B. Dalena et al.
Phys. Rev. ST Accel. Beams 15, 051006 2012.

vii   **Integrated simulation of ground motion mitigation techniques for the future compact linear collider (CLIC)**
G. Balik et al.
Nucl. Instrum. and Meth. A, 700:163-170, 2013

viii   **Analytic robust stability analysis of SVD orbit feedback systems**
J. Pfingstner and J. Snuverink
CLIC-Note-946, 2012

Contributions to the publications

The papers I, IV, V, VI and VII are concerned about the design and the testing of the linac feedback (L-FB) of the Compact Linear Collider (CLIC). The papers V, VI and VII were written jointly with Jochem Snuverink, who helped with the implementation of the simulation framework and the simulation work. The papers I and IV were written and worked out exclusively by myself. In paper I the effectiveness of the use of Kalman-filters for the L-FB is investigated, which is presented in this thesis in Sec. 3.2.2. In paper IV the final L-FB design is shown (Sec. 3.2.3), while in paper V the focus is laid on the created simulation framework (Sec. 2.4) and the simulated controller performance (Sec. 4.1). In paper VI robustness studies about the L-FB are shown (Sec. 4.2). Finally the results of the papers I, IV, V and VI are collected in paper VII, where also more details about the used models are given (Sec. 2.2).

The papers II and III are about the system identification algorithm for orbit response matrices. Both papers were exclusively written and worked out by myself. While paper III covers the overall algorithm that is presented in this thesis in Chap. 5, paper II describes the amplitude model for beam oscillations in the main linac of CLIC given in Sec. 5.3.

Contents

1. **Introduction**   1
    1.1. The Compact Linear Collider 1
        1.1.1. Physics motivation 1
        1.1.2. Accelerator complex 3
    1.2. Feasibility issues due to ground motion 6
    1.3. Luminosity preservation 9
        1.3.1. Static alignment 9
        1.3.2. Dynamic alignment 11
        1.3.3. Long-term alignment 16
    1.4. Outline 17

2. **Modelling and simulation of ground motion effects**   19
    2.1. Ground motion 19
        2.1.1. Basics of stochastic processes 19
        2.1.2. Ground motion models 24
    2.2. Luminosity and beam orbit impact of ground motion 32
        2.2.1. Basics 33
        2.2.2. Sensitivity function for beam oscillations 38
        2.2.3. Sensitivity function for Luminosity loss 41
        2.2.4. Use of the models for L-FB design and performance prediction . . 43
    2.3. Effects due to the final doublet offsets 46
        2.3.1. Beam-beam offset 46
        2.3.2. Beam size growth 48
    2.4. Integrated simulation framework 52
        2.4.1. Overview 52
        2.4.2. Individual components 52

3. **Controller design**   57
    3.1. Introduction 57
        3.1.1. Classification of accelerator feedback systems 57
        3.1.2. Literature review and historical overview 58
        3.1.3. Conclusions for CLIC 61
    3.2. Linac controller 62
        3.2.1. Problem statement and system model 62
        3.2.2. Control structure choice 67
        3.2.3. Controller optimisation 77
        3.2.4. Discussion of the novel controller design method 90
    3.3. Alternative designs for hardware cost reduction 92

Contents

    3.4. IP controller 96

4. Controller performance, imperfections and robustness studies   **99**
    4.1. Performance in presence of ground motion effects 99
    4.2. Imperfections and robustness 101
        4.2.1. BPM resolution 101
        4.2.2. Energy imperfections, dispersion filter and energy measurement . . 103
        4.2.3. Quadrupole position errors 105
        4.2.4. Other imperfections 107
    4.3. Conclusions 110

5. System identification scheme for orbit response matrices   **113**
    5.1. Introduction 113
        5.1.1. Motivation 113
        5.1.2. System identification for orbit response matrices 114
        5.1.3. Interleaved, model-supported system identification scheme 116
        5.1.4. Related work 117
    5.2. Details about the identification scheme 118
        5.2.1. Excitation unit 118
        5.2.2. Local RLS algorithm 119
        5.2.3. Phase reconstruction and merging 121
    5.3. Amplitude model for the main linac of CLIC 122
        5.3.1. Introduction 122
        5.3.2. Description of beam oscillations in the main linac of CLIC 122
        5.3.3. Calculation of the basic amplitude model 126
        5.3.4. Extensions to the basic amplitude model 127
        5.3.5. Amplitude model validation 129
    5.4. Simulations results of the identification scheme with the L-FB 130
    5.5. Conclusions 132

6. Conclusions   **133**

A. Beam physics background   **135**
    A.1. Luminosity 135
    A.2. Emittance and beam size 136
    A.3. Energy spread and dispersion 137

B. Integrated simulation framework   **139**
    B.1. Download and installation 139
    B.2. Usage and interface 139
    B.3. Parameters 141

C. Control engineering background   **145**
    C.1. $\mathcal{Z}$-transform 145
    C.2. Controller design using the frequency domain 146
        C.2.1. Standard control loop 146
        C.2.2. Loop shaping 147

      C.2.3. Bode's sensitivity integral 149
C.3. Kalman-filtering 150
C.4. Singular value decomposition 151
C.5. RLS algorithm with exponential forgetting 151

List of abbreviations

$\Delta\epsilon$	Emittance increase
$\Delta\mathcal{L}$	Luminosity loss
δ	Beam-beam offset at the interaction point
σ^*	Beam size at the interaction point, expressed in the standard deviation of the beam distribution
BDS	Beam Delivery System
BPM	Beam Position Monitor
CDF	Cumulative Distribution Function
CLIC	Compact LInear Collider
DESY	Deutsches Elektron-Synchrotron
DFT	Discrete Fourier Transform
DP	DiPole magnet
FD	Final Doublet
FF	Final Focus
ILC	International Linear Collider
IP	Interaction Point
IP-FB	Interaction Point FeedBack
IRMS	Integrated Root Mean Square
IT-FB	Intra-Train FeedBack
L-FB	Linac FeedBack
LHC	Large Hadron Collider
MIMO	Multi-Input Multi-Output
PDF	Probability Density Function
PSD	Power Spectral Density

Contents

QD0   Last quadrupole before the interaction point

QF1   Second to last quadrupole before the interaction point

RV   Random Variable

SISO Single-Input Single-Output

SLAC Stanford Linear Accelerator Center

SP   SextuPole magnet

STP   STochastic Process

STR   Self Tuning Regulator

SVD   Singular Value Decomposition

1. Introduction

1.1. The Compact Linear Collider

1.1.1. Physics motivation

A large number of particles from outer space bombard the atmosphere of our planet every day. In the first half of the 20th century, scientists discovered several until then unknown particles in the particle showers resulting from this bombardment. To study the properties of the new particles in detail, physicists try to imitate the particle collisions in the atmosphere with the help of particle accelerators. In these machines, charged particle beams are accelerated to very high energies before they are collided in a controlled way. The collisions points are surrounded by complex measurement devices called particle detectors.

In beam collisions in accelerators, new particles are created, which are very rare on our planet due to their instable nature. This apparent creation of new particles out of nothing is against our common sense, but is possible in the sub-atomic world. The mechanism can be explained by the theory of relativity and quantum field theories. In quantum field theories certain conservation laws have to be fulfilled, e.g. energy conservation. The number of the involved particles in a certain reaction does not have to be preserved though. Therefore, new particles can be created with a certain probability, if conservation laws are not violated. Also the mass of the involved particles can change. This change of particle mass can be explained by the formula $E = mc^2$ (special relativity), which states that mass is just another form of energy. At a particle collision, kinetic energy from two colliding particles can therefore be transformed into mass. Only the sum of all energies—including the energy in the form of mass—must be constant. Physicists use this mechanism to search for unknown, heavy and rare particles. Observations of such particles validate or rule out proposed theories of theoretical physicists. Not only the existence of the particles, but also their exact properties are important for this validation process.

One of the particles, which is of current research interest, is the Higgs particle and the corresponding Higgs field. The Higgs particle is the only particle of the Standard Model of particle physics (SM) which has not been observed yet. For explanation, the SM is a very powerful theory that describes three of the four known fundamental forces: the electromagnetic force; the weak nuclear force, which is responsible for e.g. radioactivity and the strong nuclear force, which holds the elements of nucleons together. Only gravity is not included in this model. The SM consists of two families of particles: ordinary elementary particles and force carrier particles. Forces acting between ordinary particles are described by the exchange of force carrier particles.

The first version of the unification of the electromagnetic and the weak nuclear force, the so called electro-weak force, predicted that all force carrier particles of the electro-

1. Introduction

weak force should be massless. This is not the case in reality though. To correctly predict the masses of these force carrier particles, spontaneous symmetry breaking and the Higgs mechanism had to be included in the theory. In the extended theory, the mass of the particles is interpreted as the strength of their interaction with the Higgs field. This mechanism predicts also the existence of a Higgs particle. To validate the theory of the Higgs mechanism, the discovery of the Higgs particle is indispensable. Recent exciting results from CERN actually conform the existence of a "Higgs-like" particle with a rest mass of about 126 GeV, but further studies are necessary to measure the detailed properties of this discovered particle.

Another topic of high interest in today's physics is super-symmetry (SUSY). SUSY is an extension of the SM. According to this theory, every particle in the SM has a super-symmetric partner, which differs from its partner in the SM only by the spin and the mass. Many theories beyond the standard model, e.g. string theories, use the idea of super-symmetry. These theories aim to include also gravity into to the SM and to unite the strong nuclear and the electro-weak force into one formalism. Super-symmetric particles could also be of interest for a phenomenon called dark matter, which is a proposed explanation of the fact that the motion of galaxies cannot be explained by their mass distribution. Star clusters behave in a way that leads to the conclusion that there should be approximately ten times as much mass in space as can be observed. This missing mass is called dark matter and one candidate is matter consisting of super-symmetric particles.

The validity of the Higgs mechanism, SUSY and other physical theories are tested at the Large Hadron Collider (LHC), which is at the publishing date of this thesis the most powerful particle accelerator. The LHC collides proton (p^+) beams with a centre of mass energy of 14 TeV (at the publishing date only 8 TeV). The creation of new particles does not occur due to the interaction of the protons itself, but due to the interactions between the fundamental particles forming the protons, which are quarks and gluons. While the energy of the proton can be determined accurately, the energy of the constituting quarks and gluons is a fluctuating stochastic process. Therefore, the collisions of the LHC have a natural energy spread, which is desirable to find unknown particles (*discovery machine*). However, to determine the detailed properties of the discovered particles accurately, the large energy spread is counterproductive. A collider that collides elementary particles would be preferable to perform such measurements (*measurement machine*). At the moment there are two main proposals for such a measurement machine: the *International Linear Collider* (ILC) (see RDR [90]) and the *Compact Linear Collider* (CLIC) (see CDR [107] and Ellis and Wilson [38]). Since this thesis is only concerned with CLIC, we will focus on this machine in the following. CLIC collides electrons (e^-) and positrons (e^+) at a centre of mass energy of 3 TeV. In addition to the advantage of better defined collision energy, the particle creation in e^-e^+-collisions shows different characteristics from the collisions of quarks and gluons, which complements the LHC measurements.

There are many important measurements that could be performed at CLIC. In case the Higgs particle exists, CLIC could reveal important additional information about it (see Battaglia et al [11] and Adli [3]). The theory predicts that the Higgs is a spin-0 particle, which could be easily verified with CLIC measurements. Such a result could verify that the newly found particle at the LHC is really the Higgs and not another particle with similar properties. Also, the interaction strengths (coupling) of other particles to the Higgs could be measured with relatively high accuracy. In the easiest proposed variant of the Higgs mechanism, these interaction strengths are strictly proportional to the mass of the particles. There are other variants proposed though and a precise knowledge of the interaction strengths is important to find the theory that describes nature best. The interaction of the Higgs particle with itself would give important new insights, by allowing to determine the exact form of the Higgs field potential. Another major research field for CLIC is SUSY (see Battaglia et al. [11] and Ellis [37]). Experiments at CLIC could complete the spectrum of SUSY particles, which is not possible at the LHC. The more accurate measurements of SUSY particle properties at CLIC are necessary to understand which of the proposed SUSY theories describes nature best. Other interesting fields of research for CLIC are less popular theories that provide alternative explanations for the questions answered by Higgs mechanism and SUSY (see Battaglia et al. [11]). Also effects of some theories including extra space dimensions could possibly be observed (see also Battaglia et al. [11]) at CLIC.

1.1.2. Accelerator complex

CLIC is a linear collider, which is planned to be installed in a 50 km long tunnel about 100 m underground. The reason for the linear form is synchrotron radiation. If the trajectory of charged particles is bent by a magnetic field, they radiate photons (synchrotron radiation). The power radiated by an ultra-relativistic particle scales as $P \propto E^4/R^2/m_0^4$ (taken from Wille [136]), where E is the particle energy, R is the bending radius and m_0 is the rest mass of the particle. Since the rest mass of an electron or a positron is orders of magnitude smaller than the mass of a proton, the emitted synchrotron radiation is much stronger for electrons and positrons, for equal particle energies. In the energy regime of CLIC, electron- and positron-beams would lose too much energy when moving in a ring of similar size to the LHC. The operation of such a ring collider would be too inefficient and since also the circumference of the collider is limited by cost issues, CLIC was chosen to be of linear structure.

The necessary linear form of CLIC implies several difficulties for the design. The major difference from a ring is that an accelerated beam can only be used once, before it is stopped by the so-called beam dumps. The beam energy is converted into heat and is lost for collisions. This is very different at a ring collider, in which a beam—once it is accelerated to its final energy—can be collided several million times. It is therefore much easier for a ring collider to run efficiently than for a linear collider, which would have an unacceptable energy consumption if it would run at the same high repetition than a ring. To reduce the power consumption, CLIC has to adapt a different beam structure as depicted in Fig. 1.1 (left). As can be seen, every 20 ms 312 beam bunches, called beam train, collide during 156 ns. This means that in the majority of the time no collisions occur. In order to run CLIC efficiently, the accelerator has to consume as little power as possible during the beam-free period. To accomplish this, the power-consuming acceleration fields are created in so called acceleration structures only during the very short time of the beam passage. Conventional creation of these fields via klystrons would be inefficient, since klystrons are optimised for continuous and not for pulsed operation. Instead the so called two beam acceleration scheme is used, where an additional electron-beam—the so called *drive beam*—is used to accelerate the *main beam*. The drive beam is

1. Introduction

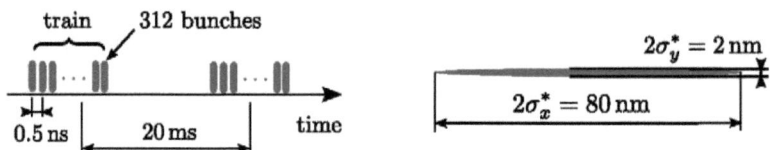

Figure 1.1.: Longitudinal (left) and transversal (right) beam structure of CLIC. (left) The CLIC beam is composed of approx. 0.3 ps long beam bunches (corresponds to $2\sigma_z = 90\,\mu m$). With a spacing of 0.5 ns, 312 such bunches are combined to one beam train (also called pulse). Consecutive trains are separated by 20 ms. (right) The transverse beam profile at the IP is close to Gaussian in both direction, with a core beam size σ_x^* and σ_y^* (fit of a Gaussian to the real beam). The beam was chosen to be flat, to lower beamstrahlung. At the beam collisions the strong electromagnetic forces between the beams accelerates the particles transversely, which forces the particles to emit synchrotron radiation (beamstrahlung). Beamstrahlung changes the energy spectrum of the colliding beams in an unwanted way. To limit this degradation, the beam sizes are chosen to be flat. This choice is beneficial, since the beamstrahlung is $\propto 1/(\sigma_x^* + \sigma_y^*)$, while the luminosity is $\propto 1/(\sigma_x^* \sigma_y^*)$.

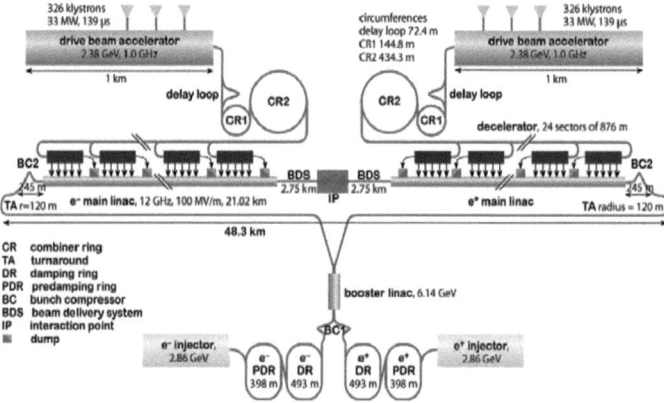

Figure 1.2.: CLIC complex. The upper part of the plot shows the drive beam complex. It consists of particle sources, efficient acceleration of the bunches with klystrons and the overlapping of the bunches with one delay loop and two combiner rings. The drive beam is then transported to the power extraction units (in blue) to power the acceleration structures of the main linac. The lower part of the plot shows the main beam complex. Electrons and positrons are created in dedicated sources. After a pre-acceleration (injector), the beam sizes are shrunk in the so-called damping and pre-damping rings to the necessary size. The main linacs accelerate the electron and positron beams to their final energy, before it is conditioned in the beam delivery system for the collision at the interaction point.

1.1. The Compact Linear Collider

accelerated via klystrons with high efficiency, before it is shortened and intensified at the same time, by overlapping the beam bunches several times in a delay loop (Schulte [110]) and two combiner rings (Corsini and Delahaye [28]). The shortened drive beam is then transported (Adli [3]) to the power extraction units (PETs) (Syratchev [131]), which are located beside the accelerating structures of the main linac. In the PETs the drive beam is decelerated, which results in the creation of electromagnetic waves used to power the accelerating structures shortly before the arrival of the main beam. The CLIC accelerating structures (Grudiev and Wuensch [49]) are not super-conducting, since otherwise the energy transfer between the PETs and the structures could not be carried out fast enough. Another advantage of the normal-conducting structures is that the acceleration gradients are not fundamentally limited as for super-conducting structures. CLIC can therefore be built more compact (*Compact Linear Collider*). The achievable accelerating voltage gradient is limited by the appearance of field breakdowns, due to mechanical fabrication tolerances and the heating of the structure by the accelerating fields. The maximal allowable breakdown rate and the power consumption limit the necessary time between beam trains.

As discussed, the beam collision rate is much lower at CLIC than at the LHC. To be able to create enough particle collisions the beam collisions have to be more efficient than at the LHC. To understand how this can be achieved, it is necessary to introduce the associated quantity that is called *luminosity* $\mathcal{L}$. The luminosity is direct proportional to the collision rate and therefore, beside the collision energy, the most important parameter of a particle collider. It scales as $\mathcal{L} \propto N^2/\sigma_x^*/\sigma_y^*$, where N is the number of particles in the beam, σ_x^* is the horizontal and σ_y^* the vertical beam size (corresponds to the standard deviation of a Gaussian curve fitted to the particle position histogram). The number of particles in the beam is limited mainly by the electric fields induced by the beam in the accelerating structures (called wake fields). Too strong wake fields would destroy the beam quality during acceleration and therefore a higher number of particles can only be used to a certain extent to increase the luminosity. Thus, the transversal beam sizes σ_x^* and σ_y^* have to be lowered (see Fig. 1.1 (right)), to achieve the luminosity goal. The most important design parameters of CLIC are summarised in Tab. 1.1.

The complete CLIC complex that is used to create the necessary beams is depicted and explained in Fig. 1.2. A detailed explanation of all elements can be found in the CDR [107] and in CLIC 2008 Parameters [15]. This work is concerned with the two main linacs (Schulte [109]), in which the beams are accelerated to the collision energy of 1.5 TeV per beam, the two beam delivery systems (BDSs) and the interaction point (IP), both explained in Tomás et al. [132]. In the BDSs the beams are pre-conditioned for the interactions at the IP. Particles that deviate too strongly from the nominal beam position and energy are removed by so-called collimators. Additionally, the beams are focused to the very small beam sizes at the IP in the so-called final focus systems.

The accelerator community has identified several especially challenging issues for the design of CLIC. These so called *feasibility issues* are not only technological demanding, but would also lead to an unacceptable performance reduction if the according specifications could not be met. The feasibility issues are grouped in the four categories

- two beam acceleration scheme,
- ultra-low emittances and beam sizes,

1. Introduction

Centre of mass energy	E	3 TeV
Total/peak (1 %) luminosity	$\mathcal{L}/\mathcal{L}_{1\%}$	$5.9/2.0 \times 10^{34}\,\mathrm{cm^{-2}\,s^{-1}}$
Hor./vert. beam size at IP	$\sigma^*_{x/y}$	40/1 nm
Hor./vert. norm. emittance at IP	$\epsilon^*_{x/y}$	660/20 nm rad
Nr. of particles per bunch	N	3.72×10^9
Repetition rate	f_R	50 Hz
Nr. bunches per beam train	N_b	312
Bunch interval	Δ_b	0.5 ns
RF gradient	E_a	100 MV/m
Total power consumption	P_{total}	560 MW

Table 1.1.: Most important design parameters of CLIC. Parameters in the table that have not already been explained are introduced in App. A.

- operation and machine protection system and
- detector.

In this thesis, essential contributions to the feasibility issues of ultra-low emittances and beam sizes will be presented. The focus will be on the mitigation of ground motion effects, which are introduced in the next section. For an overview of the current status of the CLIC project and the future plans, please refer to Stapnes and Schulte [126].

1.2. Feasibility issues due to ground motion

Ground motion is a severe problem for CLIC. The movement of the tunnel floor displaces accelerator elements from their initial position, which causes performance reduction. Without countermeasures, ground motion would degrade the luminosity to an unacceptable level even on a train-to-train basis. In this section we illustrate the basic principles of performance degradation due to ground motion. These principles will be quantified in Chap. 2, where detailed models of ground motion and its effect on beam parameters are given. The reader is assumed to be familiar with the most basic terms of beam physics, which are covered briefly in App. A.

The luminosity is lowered by ground motion in two ways: IP beam size growth $\Delta\sigma^*$ and beam-beam offset δ (see Fig. 1.3). To relate these quantities to the luminosity loss, δ and $\Delta\sigma^*$ have to be normalised to the IP nominal beam size σ^*. The σ^* of CLIC is only 1 nm in vertical and 40 nm in the horizontal direction and thus small compared to other machines. At the LHC e.g., the design value of σ^*_y is 16.7 μm and hence more than four orders of magnitude larger than at CLIC. The most powerful linear electron-positron collider which has ever been in operation—the SLC—reached a σ^*_y of 0.55 μm and even the competitor of CLIC, the ILC, only aims for a σ^*_y of 5.7 nm. Thus, already small δ and $\Delta\sigma^*$—which are negligible for other accelerators—cause significant luminosity loss

1.2. Feasibility issues due to ground motion

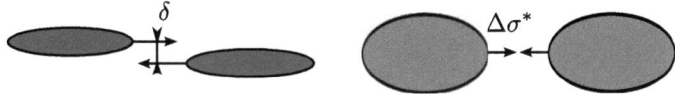

Figure 1.3.: The luminosity is degraded by two effects: the beam-beam offset δ (left) and the growth of the IP beam size $\Delta\sigma^*$ (right).

for CLIC. Additionally, the preservation of the beam quality along the main linac is especially difficult for CLIC. To quantify the beam quality the term emittance is usually used instead of the beam size. The reason for this is that the beam size depends also on the strength of the magnets focusing the beam. For example, even a beam with very good quality (emittance) can be relatively large if the focusing magnets are weak. By using the term emittance the effects of the beam quality and the magnet system are separated. Coming back, the reason for the difficulty of emittance preservation in the main linac is the large energy spread of the beam in the main linac. This large energy spread (energy deviation of the individual particles from the average beam energy, see App. A.3) is necessary for a technique called BNS damping (see Balakin, Novpkhatsky and Smirnov [7]), which ensures that the beam quality is not destroyed by the strong wake fields in the accelerating structures. However, in the case of ground motion the high energy spread is counter-productive, due to the effect of filamentation, which will be explained in a moment.

CLIC is most sensitive to the displacement of quadrupole magnets. Simulations in Schulte and Tomás [114] show that the tolerance for 1% luminosity loss for a random (Gaussian) displacement of the quadrupoles in the main linac is as small as 1.8 nm. The tolerance for 1% luminosity loss for the position of the two last magnets of the BDS is even only 0.1 nm (see Schulte [112]). To illustrate the scale of the problem: the covalent radius of a sulphur atom is 0.104 nm. Even at sites with low ground motion, the differential motion exceed these limits rapidly. However, the estimates above are made for uncorrelated motion of the quadrupoles. In reality the ground motion source that causes the strongest excitation in the frequency range of interest is strongly spatially correlated over large distances. Such spatially smooth misalignments, even though with high amplitudes, are not as harmful for the beam as random misalignments. On the other hand, ground motion with a wavelength close to the natural beam oscillation frequency of the lattice (betatron wave length) is most harmful for the beam, since the excitation acts in a resonant manner. For such resonant excitations, the misalignment tolerances are much smaller than for uncorrelated excitation. Note that since the form of the misalignment matters, ground motion effects have not only a temporal, but also a spatial dependence. This fact will be fruitfully deployed by the ground motion mitigation methods developed in this thesis.

The main mechanisms by which ground motion creates beam-beam offset δ and beam size growth $\Delta\sigma$ are the following. If a quadrupole is misaligned, the nominal beam passes through it with an offset Δy. Due to this offset, the beam is not only exposed to the usual quadrupole field, but also to a dipole field, which exerts an additional kick on the beam. The kick deflects the beam transversally, which causes so called betatron oscillations (see Fig. 1.4 (left)). These betatron oscillations propagate to the IP and

1. Introduction

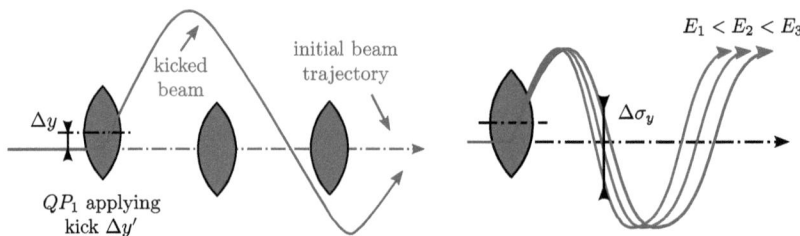

Figure 1.4.: Luminosity degradation via ground motion. (left) Misaligned quadrupoles (blue lenses) apply transverse kicks to the beam, which results in (betatron) oscillations along the beam line. The amplitude of such oscillations is proportional to the kick strength $\Delta y' = -\Delta y/f$, where f is the focal length of the quadrupole. The form (relative amplitude and phase) of the oscillation is determined by the beam line design. (right) Due to a dipole kick from a quadrupole (blue lens), a beam modelled by three particles with the energies E_1, E_2 and E_3 starts to oscillate. The particles are initially in phase, but become increasingly non-coherent over time, due to the energy difference. The beam emittance grows.

cause the beam-beam offset δ. While travelling to the IP the oscillations also lower the beam quality via an effect called filamentation, which leads to an emittance increase and consequently to a beam size growth at the IP $\Delta\sigma_y^*$.

To understand filamentation we have to distinguish between beam and particle motion. The beam motion is the movement of the average position of the particles forming the beam. Therefore, even if a beam is on its nominal trajectory, the individual particles still oscillate with random phases. The form of these so called betatron oscillations depends on the accelerator design, which can be characterised by the beta function $\beta(s)$ ($\propto$ to the square of the oscillation amplitude) and the phase advance $\phi(s)$ (see e.g. Wiedemann [135]). But $\beta(s)$ and $\phi(s)$ also depend on the particle energy. This can be easily understood by the fact that the same magnets bend a particle with higher energy less than a particle with lower energy. If a beam experiences a kick by a misaligned quadrupole, a second motion is superimposed to each particle, additionally to the betatron motion. Contrary to the betatron motion the particle motion due to the kick has the same phase for all particles. Thus, the beam oscillates as a whole. Due to the natural energy spread of the particles in a beam, particles with higher energies oscillate slower than the ones with lower energies. The motion of the particles becomes increasingly non-coherent over time (see Fig. 1.4 (right)). The projected emittance of the beam grows, while the beam oscillations are damped at the same time. Since the energy spread is larger in the main linac than in the BDS, the filamentation effect is more severe in the former one. In Sec. 5.3 the effect of filamentation will be dealt with in a more quantitative way. An amplitude model for the beam oscillations in the main linac with filamentation will be derived (also published in Pfingstner et al. [84]).

Beside the effects of beam oscillations and filamentation, ground motion also results in secondary effects on the luminosity. These effects occur mainly in the last section of the BDS, the final focus system (FF). In the FF especially strong quadrupoles and nonlinear sextupole magnets are used to focus the beam to the needed spot size at the IP. Misalignments of these elements lead to dispersion (dependence of the transversal particle position on the energy), waist shifts (longitudinal shifts of the focal point at the IP), uncorrected geometric aberrations (see Zimmermann [138]) and coupling between the horizontal and the vertical plane. These effects lead to a beam size growth at the IP. Since the mentioned dispersion effect will be taken into account for the feedback design in this thesis, a model of the beam size growth due to dispersion will be developed in Sec. 2.3.2.

It should also be mentioned that the misalignment of magnets is not only caused by ground motion. Also other vibrations, not transmitted via the ground, play a role. Examples for these disturbances are the cooling water in magnets, air flow of the tunnel cooling system, thermal drifts and sound waves. There is some work done on these topics, but the knowledge about ground motion is further advanced, since vibration transmission through the floor is considered the more severe problem. In this work only ground motion is considered.

1.3. Luminosity preservation

The problems arising from misaligned accelerator elements are counteracted at CLIC with the help of several mitigation methods. These methods can be divided by the misalignment type they are intended to cure: static, dynamic and long-term alignment. The focus of this thesis will be the dynamic alignment methods.

1.3.1. Static alignment

When an accelerator is constructed, the individual components are mounted on girders and are pre-aligned mechanically to a precision of about 0.1 mm. This tolerance is not sufficient for the needs of CLIC and has to be improved by active pre-alignment and beam-based alignment methods.

At the active pre-alignment procedure, the girders, on which the accelerator components are mounted, are aligned according to a reference system. At the current baseline, this reference system consists of a network of 200 m long stretched wires (see Touzé [133]). As an alternative, studies of laser reference system (LAMBDA project) have been started (see Lackner et al. [66]). The girders are equipped with sensors that can measure the distance from the wire to the girder. These measurements are used to re-align the girders and the quadrupoles, which are separately supported. For the girder re-alignment three degrees of freedom linear actuators are used, while for the quadrupoles five degree of freedom cam movers (positioning system based on the rotation of eccentric shafts) have been designed. The movers have to be designed as stiff as possible in order to not amplify ground motion via excitation of mechanical resonances of the system. When calculating the overall achievable alignment tolerance, also the alignment tolerances of the elements with respect to the girder reference system (fiducialisation) and the interconnection of the girders have to be taken into account. Considering these parameters the quadrupoles are assumed to be aligned better than to a standard deviation of 17 μm. Over longer distances the misalignment tolerances (with respect to a straight line) are much larger,

1. Introduction

but these smooth misalignments are not that harmful to the beam as the ones with less correlation.

Even though the capability of aligning massive objects as girders and quadrupoles to a precision of 17 μm is impressive, it is still not sufficient for CLIC. A set of additional beam-based alignment methods has to be applied. These methods differ in some aspects between the main linac and the BDS. For the main linac, the beam-based alignment is a mature procedure and the following explanations, therefore, mainly refer to them. But also for the beam-based alignment and the luminosity tuning of the BDS, significant improvements have been achieved recently (see Latina and Raimondi [68] and Dalena et al. [29]).

When a beam travels the first time through the pre-aligned accelerator the still misaligned quadrupoles cause the beam to oscillate strongly. This causes a large emittance growth due to filamentation. To resolve this problem, the beam oscillations are measured with beam positioning monitors (BPMs), which are mounted to the quadrupoles. The position of each quadrupole is now varied one after the other, such that the beam is centred in the down-stream BPM. When this procedure—called 1-to-1 steering—is finished, the emittance growth is strongly decreased, and the beam is centred in all BPMs. As an alternative to vary each quadrupole separately, the accelerator can also be subdivided into sections, in which the BPM readings are minimised simultaneously by solving a system of equations (few-to-few steering).

Even though the passing beam is now centred in all BPMs, there remains the problem that the BPMs are not perfectly aligned to the magnetic centres of the according quadrupoles. To reduce the remaining dipole kicks a sophisticated technique called dispersion free steering (DFS) is used (see Raubenheimer and Ruth [99]). The basic principle is the following. If a beam passes exactly through the centre of a quadrupole it experiences no dipole kick and therefore no deflection. Therefore, an energy variation of the beam does not change its trajectory through the down-stream BPMs. If on the other hand a beam is offset in a quadrupole, the down-stream motion will be dependent on the beam energy, since a beam with high energy will be less deflected by the same kick as one with lower energy. This fact can now be exploited. By probing the accelerator with two beams of different energy, the according BPM readings can be used to find quadrupole positions for which the difference of the BPM reading of beams with different energies is minimised. The quadrupole kicks are strongly reduced and therefore also the emittance growth. Beside DFS there are other similar alignment methods proposed, such as ballistic alignment and kick minimisation. An overview is given in Raubenheimer and Tenenbaum [100].

Since the misalignments after the pre-alignment are relatively large, not only offsets of quadrupoles but also accelerating structure offsets matter. A beam that passes with an offset through a structure induces wake fields, which increase the emittance. To cope with this problem, a so called wake field monitor is installed in every structure to measure the beam offset. The wake field kick of the structures of each individual girder is minimised on average, by moving the average position of the structures onto the beam trajectory. Only the average kick is minimised, since the structures of one girder can only be moved together. To remove remaining wake field kicks so called emittance bumps can be used (see Eliasson [36]). Dedicated structures are moved by an optimisation algorithm such that the beam size at the end of the linac is minimised.

1.3. Luminosity preservation

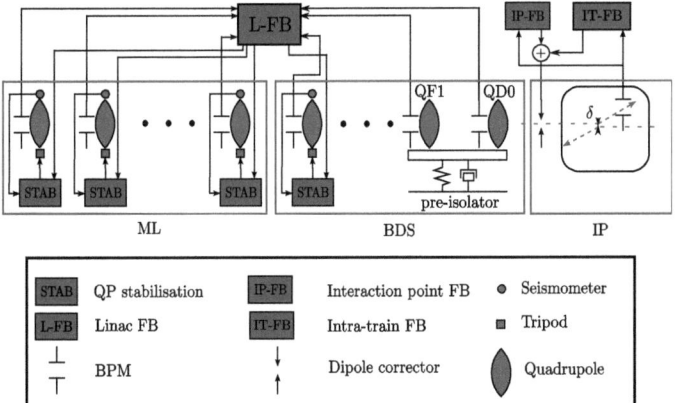

Figure 1.5.: Overview of the ground motion mitigation methods of CLIC. While the quadrupole stabilisation is a local method, the linac feedback (L-FB) acts globally. The pre-isolator (mass-spring-damper system) stabilises the last two quadrupoles of the BDS called QF1 and QD0. Additionally, the interaction point feedback (IP-FB) and the intra-train feedback (IT-FB) reduce the beam-beam offset δ.

Such emittance bumps can also be deployed to compensate dispersion. At the moment the emittance bumps are not in the baseline design of CLIC and are kept as a reserve.

1.3.2. Dynamic alignment

After the accelerator is statically aligned, a second misalignment problem arises. Ground motion misaligns the statically aligned accelerator components gradually, which causes luminosity decrease. This decrease is caused by beam-beam offset due to beam oscillations and beam size growth mainly caused by filamentation but also by dispersive effects. Four dynamic alignment methods—plus one, which is kept as a reserve—are used at CLIC to cure these effects. These methods are intended to keep the luminosity loss for several minutes within the assigned budget. After about 20 minutes also secondary effects start to become important, which are mitigated with dedicated long-term methods (see Sec. 1.3.3). An overview of the dynamic mitigation methods is given in Fig. 1.5. The individual systems are explained in detail in this section.

1.3.2.1. Linac feedback

The linac feedback (L-FB) is intended to control the beam orbit all along the main linac and BDS of CLIC. The reader should not be confused by the name, which could be interpreted as a feedback system only for the main linac. There are two independent L-FBs used, one for the electron and one for the positron part of CLIC. The primary task within the orbit control problem is to suppress beam oscillations caused by quadrupoles, which are displaced by ground motion or other disturbances. As explained in more detail

1. Introduction

in Sec. 1.2, these oscillations cause a growth of the beam emittance and can also lead to beam-beam offsets at the interaction point (IP).

In order to counteract these effects, the beam oscillations are measured with beam position monitors (BPMs), which are distributed along the beam line. Note that only the average position of each train is measured and therefore the L-FB only acts on a train-to-train basis. The BPM measurements are used by a control algorithm to calculate corrector actuations that aim to steer the next beam train onto the target orbit. The target orbit can either be the centre of the BPMs, or some other orbit in case of an intended offset. At CLIC the target orbit in the vertical direction is the centre of the BPMs, while in the horizontal direction the target orbit is non-zero in the BDS. Note that by changing the reference orbit, beam bumps can be created.

The BPMs have in the current baseline an accuracy of $5\,\mu$m and a resolution of 50 nm in main linac and BDS. Especially the resolution is a demanding requirement considering the high number of devices: 2009 in each of the main linacs, 113 in each of the BDSs, which makes all together 4244 BPMs. Each device measures in vertical and horizontal direction. For sake of completeness, we want to mention that the type of BPM in the current baseline choice is a so called resonant cavity BPM, and a choke-mode cavity BPM is in discussion as an alternative.

Also the 2104 actuators per beam line are designed to correct the beam oscillations in vertical and horizontal direction. There are two possible actuator choices. The first one is based on the transversal displacement of quadrupoles. Such a displaced quadrupole acts on the beam as the original quadrupole plus a dipole. The strength of the dipole kick is proportional to the strength of the quadrupole and the amount of displacement. For the displacement of the quadrupoles the stabilisation system (see Sec. 1.3.2.4) would be used. Additionally to the main task of stabilising the quadrupoles, the stabilisation system would also take over the task of a positioning system. In the current design, the positioning resolution has to be in the order of 0.25 nm in the BDS and 0.5 nm in the main linac. These tolerances would keep the luminosity decrease below 1 %. The second actuator option is the use of dipole corrector magnets, which are embedded into the quadrupoles. The advantage of this option is that the task of stabilisation and positioning can be split up and the demands on the stabilisation system are relaxed. On the other hand the cost increases, since the stabilisation system will still have to be put in place as well.

The control algorithm, connecting the sensor readings and the actuators, has to take into account the special nature of the accelerator system. A first characteristic is the huge dimension of the control problem with 2122 sensors and 2104 actuators. Simplifications in the control algorithm structure are advisable. The next characteristic is that the system is an intrinsically discrete one. The beam trains of CLIC are separated by 20 ms. This sampling rate limits the maximal resolvable frequency, due to the sampling theorem, to 25 Hz. Due to principle limitations of feedback control (see App. C.2.2 and C.2.3), the L-FB is thus only capable of suppressing ground motion effects below about 1 to 4 Hz. Higher frequencies will be amplified. Finally, the L-FB design has to take into account the spectrum of ground motion and imperfections, corrector dynamics and robustness issues. Especially the suppression of BPM noise will turn out to be a difficult task. The design of the L-FB is one of the main achievements of this thesis (see Chap. 3).

An open issue, which is not sufficiently dealt with up to now, is the communication

1.3. Luminosity preservation

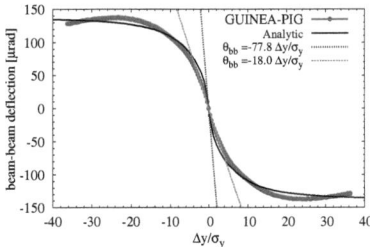

Figure 1.6.: Beam-beam deflection curve. The angle of deflection of the beams at the IP is a function of the relative beam-beam offset δ_y/σ_y^*. In this plot taken from Resta-Lopez, Burrows and Christian [103] the notation $\Delta y/\sigma_y$ instead of δ_y/σ_y^* is used. The linear approximations depend on the offset range of interest.

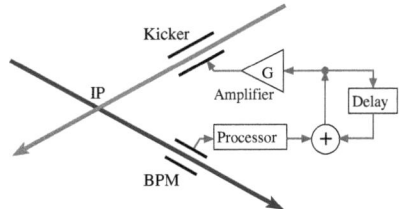

Figure 1.7.: Structure of the IT-FB taken from Resta-Lopez, Burrows and Christian [103]. The bunches travel from the kicker to the IP, receive a deflection due to beam-beam offsets, which can be measured in the post-collision BPM. This offset is corrected by a control algorithm for the following bunches. For speed reasons the FB algorithm has to be kept simple.

network needed for the L-FB. In each time step, all BPM readings have to be transmitted to a central computer, where the control algorithm calculates actuations, which have to be transmitted to the actuators. This all has to happen in real time within 20 ms. The difficulty of this task becomes obvious, when this delay time is compared with the one of the state-of-the-art system of the LHC, which is 100 ms (see Steinhagen [129]).

1.3.2.2. IP feedback

When the electron and positron beams collide at the IP, the electro-magnetic fields of the beams influence each other. As a result the beams are deflected, if they are not exactly centred with respect to each other. The deflection angle is a function of the relative beam-beam offset δ/σ^* as can be seen in Fig. 1.6. As the beams move away from the collision point, the received beam-beam kick results in transversal offsets of the beams from their nominal trajectories. This offset can be measured with a BPM 3 m down-stream of the IP in the post-collision line. This BPM signal can be used to calculate the according beam-beam offset δ, by using the beam deflection curve or its linear approximations in Fig. 1.6.

Since δ is a very important quantity for the luminosity performance a dedicated feedback system—the so called interaction point feedback (IP-FB)—has been put in place to reduce the beam-beam offset. The post-collision line BPM measurement is used in the first step to calculate the according beam-beam offset. This signal is the input for a linear, single-input, single-output control algorithm, which calculates actuator setting for a kicker magnet. The actuator settings are transmitted to two dipole corrector magnets, which are located 3 m in front of the IP on the electron and positron side respectively. Each of the two kickers applies half of the necessary kick to steer the next beam trains onto each other.

13

1. Introduction

Similar as the L-FB, the IP-FB only works on a train to train basis. Therefore, also the IP-FB can only reduce δ for frequency components below 1 to 4 Hz. Higher frequencies are amplified by the controller. In contrast to the main linac BPM readings the beam-beam offset data of the IP-FB have a very low noise level of only 10-20 pm for the beam-beam offset, depending on the final BPM choice and the range of beam-beam offset. This extremely high resolution can be reached, because the beam can build up a large transverse offset as it travels the 3 m to the post-collision line BPM. Due to the higher quality of the sensor signal, the direct measurement of the relevant quantity δ and the much smaller system dimension (one input and one output) the IP-FB can be stronger optimised than the L-FB. A collaboration of the institutes SYMME and LAPP from the Université de Savoie designs the IP-FB with the support of CERN (see Caron, Balik and Brunetti [20], Balik et al. [8] and [9]). In Sec. 3.4 a less optimised, but still efficient design is presented, which is very useful if fast changes are necessary. Both designs handle the L-FB and the IP-FB as independent systems, which is a simplification. In reality also the L-FB will influence the beam-beam offset. Since the two feedback systems do not exchange any data, their effective disturbance rejection frequency responses are multiplied.

1.3.2.3. Intra-train feedback

The intra-train feedback (IT-FB) is very similar to the IP-FB. It has the same structure and task as the IP-FB and uses also the same sensor. As an actuator a different kicker has to be used, which addresses the higher dynamic requirements and the lower necessary actuation range. The IT-FB uses very fast electronics for the BPM readout, the controller hardware and the kicker amplifiers. Therefore, it can act within a bunch train of only 156 ns. The structure of the system is depicted in Fig. 1.7. The delay time of the current IT-FB is only 37 ns, from which 20 ns are already due to the beam travel time from the kicker to the IP and further to the BPM. To save time only one kicker is used contrary to the IP-FB.

The IT-FB described above was first designed for the ILC. Due to the large bunch separation of 369 ns at ILC (compared to the 5 ns of CLIC), the IT-FB is very efficient. Hence, the IP-FB is the essential methods for ground motion mitigation at the ILC. The design envolved from the three analogue control circuits FONT1, FONT2 and FONT3 (see Burrows et al. [19]) to the current baseline FONT4, which uses digital electronics. Since CLIC has higher demands on the speed of the electronics, FONT3 (see Burrows et al. [18]) is used, since it is an analogue and therefore fast system. Still, the system is less effective than at the ILC, since the feedback algorithm can only update the kicker settings five times during one beam train. Nevertheless, simulations in Resta-Lopez, Burrows and Christian [103] show that the tolerances for the offsets of the final doublet magnets can be relaxed by a factor of two. The IT-FB is kept as a reserve in the current baseline design of CLIC.

1.3.2.4. Stabilisation system

The L-FB suppresses ground motion effects below 1 to 4 Hz efficiently. Even though the ground motion spectrum drops quickly for higher frequencies, the remaining frequency components up to 85 Hz decrease the luminosity too strongly to be neglected. These higher frequency disturbances result mainly in beam-beam offset, but also in emittance growth. A quadrupole stabilisation system has been designed to address the problem. The idea of this system is to stabilise the quadrupoles of the main linac and BDS mechanically, without considering the particle beam. It is assumed that if the quadrupoles are kept stable, also the beam receives no dipole kicks. There are attempts to verify this assumption by measurements (Gasior et al. [45]). Each quadrupole of the main linac and BDS (apart from the two final doublet quadrupoles; see pre-isolator part of this section) is equipped with such a system. All stabilisation systems work independent of each other.

The passive and active damping of mechanical vibrations—vibration control—is a well studied problem (Preumont and Seto [94]). There are numerous applications in industry and science as semiconductor lithography, nanotechnology, interferometry, large-scale telescopes and gravitational wave detectors. The vibration control specifications for accelerators, differ from all applications above. Since no solution could be adopted directly, the linear accelerator community started activities two decades ago. In 1996, Montag [75] built a stabilisation system, which was installed at the Deutsches Elektronen-Synchrotron (DESY). He could show a quadrupole stabilisation to the level of an integrated root mean square (IRMS, for a definition see Eq. (2.8)) of 25 nm for frequencies above 2 Hz, even in very noisy environment. In 2004, Redaelli [101] used a commercially available stabilisation system (Statics2000 by TMC) to show a quadrupole motion reduction to 1 nm above 2 Hz. The current baseline system of CLIC has been entirely designed and built at CERN (see Collette et al. [24]). A vibrations reduction to a level of 0.8 nm above 2 Hz was reported in Collette et al [26].

The CERN stabilisation system uses a seismometer from Guralp Systems to measure the quadrupole vibrations in the frequency range from 0.03 to 150 Hz. These measurements are used by a controller that is based on the decoupling of the different inputs and outputs (decoupling controller). The individual controller for the decoupled channels integrate the sensor data and contain a high pass filter to cut away slow drifts. Additionally, lead elements are used to improve the controller stability. The actuator used, is a complex positioning system called tripod (see Fig. 1.8 (left)), which is also used by the L-FB as an actuator.

A full mock-up of of the stabilisation system with quadrupole is not yet available. However, experiments on a scaled-down version (only two legs) confirmed the validity of the theoretical frequency responses in Fig. 1.8 (right). Simulations presented in this thesis showed in which way the baseline frequency response (version 1) could be improved. As a result the stabilisation group proposed the optimised frequency response (version 2, blue) (see Sec. 4.1 and Janssens et al. [59]). The main difference of the designs is that version 1 uses a seismometer and version 2 a geophone as sensor. An introduction to the principles and differences of seismometers and geophones is given in Collette et al. [27].

1.3.2.5. Pre-isolator

The beam-beam offset is most sensitive to offsets of the last two quadrupoles before the IP called QF1 and QD0, which form the *final doublet* (FD). Even the stabilisation system version 1 cannot sufficiently mitigate the ground motion components above 1 Hz. The

1. Introduction

 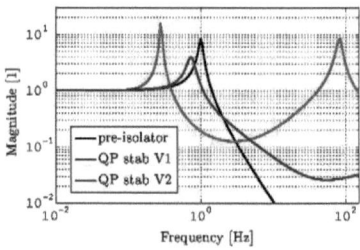

Figure 1.8.: (left) The tripod is similar to a Steward platform, and consists of a set of legs, which support the quadrupole. The length of the legs can be adapted via piezo-electric actuators from PI Physics Instruments. The quadrupole can be moved in six degrees of freedom. Picture courtesy of Artoos. (right) Absolute value of the frequency responses of the pre-isolator (black), the quadrupole stabilisation version 1 with the CMG-6T seismometer of Guralp (red) and the quadrupole stabilisation version 2 with a proposed geophone sensor (blue).

solution proposed by Gaddi [43] is a huge mass-spring-damper system, called pre-isolator. It consists of a concrete block with a weight of 110 tons supported by 10 pneumatic vibration isolators (see Fig. 1.9). The FD elements are placed on top of the concrete block. The pre-isolator acts as a very effective low pass filter (see Fig. 1.8 (right)) between the ground motion and the FD. The frequency response plotted corresponds to a point-like model of the pre-isolator. In Sec. 2.3.1 a more detailed model will be presented, in which also the tilt mode of the pre-isolator and its effect on the beam-beam offset is considered. Due to its large dimensions, the pre-isolator combines the advantages of a low cut-off frequency and robustness against disturbances acting directly in the concrete block (in contrast to ground motion, which acts on the pneumatic vibration isolators).

1.3.3. Long-term alignment

The luminosity loss $\Delta\mathcal{L}_{total}$ caused by ground motion that remains after dynamic alignment methods are applied can be split up into the three parts

$$\Delta\mathcal{L}_{total} = \Delta\mathcal{L}_{uncorr} + \Delta\mathcal{L}_{noise} + \Delta\mathcal{L}_{residual}. \tag{1.1}$$

The component $\Delta\mathcal{L}_{uncorr}$ corresponds to luminosity loss, which could in principle be cured by the dynamic alignment. The reasons for the sub-optimal operation of the dynamic alignment methods are general limitations of feedback control (stability issues, sampling time limitation) and practical limitations (sensor and actuator performance). The second component $\Delta\mathcal{L}_{noise}$ is due to the sensor noise that is introduced into the system by the feedbacks. The last component $\Delta\mathcal{L}_{residual}$ originates from effects that can in principle not be cured by the dynamic alignment. Even if the dynamic alignment would manage to steer the beams into the centre of the BPMs and exactly onto each other, there would be a residual luminosity loss. This loss is due to long-term misalignments,

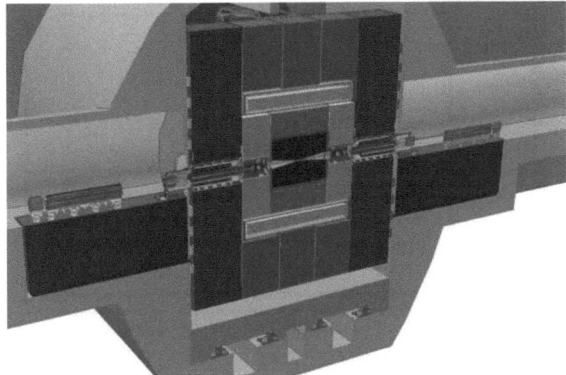

Figure 1.9.: Integration of the pre-isolator system (green) at the end of the CLIC tunnel just in front of the detector (muon chamber in red). While the quadrupole QF1 is still fully in the tunnel, the last quadrupole QD0 reaches into the detector. Picture courtesy of F. Ramos.

which are similar to the static misalignments. To cure these long-term misalignments, long-term alignment methods are foreseen. The major difference between long-term alignment methods and static alignment methods is that the long-term alignment has to work on-line (while nominal accelerator operation), which complicates the task. On the other hand long-term alignment methods can start already from a properly statically aligned machine and have to cure the slow deviation from this setup.

The work on the long-term alignment has just begun. First results are shown in Fig. 1.10 and are published in CDR [107]. The plot shows that the main luminosity lost after one hour is due to misalignments of the sextupole magnets of the final focus system. Over longer time scales also other effects are assumed to matter. As an example, the misalignments of the BPMs in the main linac lead to a change of the reference orbit with respect to the one found by dispersion free steering. This results in dipole kicks from the quadrupoles and ultimately to an emittance growth.

1.4. Outline

This thesis is about the design and validation of methods for the dynamic alignment of the main linac, BDS and IP of CLIC. The work is split up into three parts. In Chap. 2, models of ground motion and models of beam parameter variation due to ground motion are presented. Some beam physics basics, necessary to understand the material of this chapter, are given in App. A. Since all of the presented models are simplifications, the final system performance has to be verified with the help of full-scale simulations. For this task a simulation framework was set up, which is also explained in Chap. 2 and App. B.

Utilising the models from Chap. 2, the linac feedback is designed in Chap. 3. Different

1. Introduction

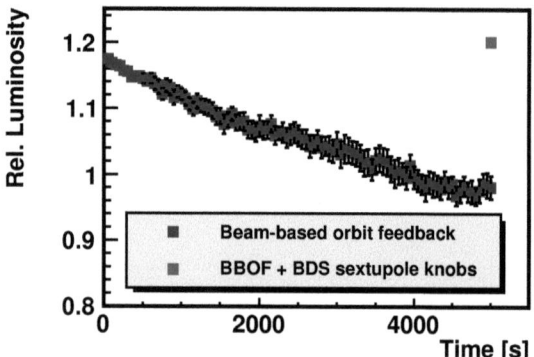

Figure 1.10.: Luminosity normalised to the target luminosity due to long-term ground motion effects. The simulation results are obtained by misaligning CLIC with ATL motion corresponding to the time axis of the plot. The dynamic mitigation methods were applied until the luminosity converged. The recovered luminosity and the error bars are plotted in blue. The relative luminosity does not start at 1 since a luminosity budget is assigned for the task of ground motion mitigation. If in addition to the dynamic alignment luminosity tuning bumps—which move the sextupoles of the FF—are used, all of the luminosity can recovered. Picture courtesy of B. Dalena and J. Snuverink.

feedback structures are compared to find the most appropriate one. After the feedback structure has been defined, the free parameters of the linac controller are calculated with a model-based approach. Two smaller studies are also presented in Chap. 3, which impact not only the linac feedback, but also the stabilisation system and the IP feedback system. Firstly, cost reduction options for the stabilisation system are investigated. Secondly, an alternative, simpler IP feedback is proposed, which is especially useful whenever fast changes of the control algorithm are necessary. For people not familiar with techniques from control engineering, supporting material is given in App. C. In Chap. 4 the simulation framework from Chap. 2 is used to validate the linac and IP feedback designed in Chap. 4. Studies about imperfections and robustness are also included.

Since the behaviour of the main linac may change significantly, mainly due to variations of the voltage gradients in the acceleration structures, a system identification algorithm is designed in Chap. 5. The classical recursive least squares algorithm is adapted to the accelerator environment of CLIC. For this reason, the algorithm has to be supported with a derived amplitude model of the main linac, which takes into account the effect of filamentation. The resulting system estimates are an important input for a variety of applications as e.g. the L-FB, beam-based alignment and diagnostics and error detection tools.

2. Modelling and simulation of ground motion effects

This chapter is concerned with the modelling of the effects of ground motion on beam parameters of a linear accelerator. In Sec. 2.1, relevant knowledge about ground motion is reviewed. Based on this material, Sec. 2.2 and 2.3 show how ground motion models can be used to predict the impact on the luminosity and the beam orbit. While Sec. 2.2 deals with models for the main linac and the BDS, Sec. 2.3 covers effects specific for the final doublet. Methods from literature will be reviewed, adapted and extended to create models of the beam parameter variation used for the controller design in Chap. 3 of this thesis. Finally, in Sec. 2.4 a developed full-scale simulation framework will be presented that allows to investigate many different ground motion related effects in an integrated fashion. With this framework also the performance of different ground motion mitigation methods can be studied.

2.1. Ground motion

In this section, fundamental information about ground motion (relevant for accelerator science) are presented. Since ground motion is not fully predictable, it has to be described mathematically as a stochastic process. Methods to quantify the properties of such stochastic processes are reviewed in Sec. 2.1.1. With the help of these tools, commonly used ground motion models can be introduced in Sec. 2.1.2. The two main models presented are the *ATL law* and models based on *power spectral densities* (PSDs). Also, numerical procedures to create realistic ground motion data for simulations are described in this section.

2.1.1. Basics of stochastic processes

The topic of *stochastic processes* (STP) is rich on theory and applications. Since the subject fills libraries, it is clear that only a brief overview of selected topics can be given here. The focus will be put on the methods used in this work, especially the *power spectral density function* (PSD) and *sensitivity functions*. The material presented in this section originates from Papoulis and Pillai [80], Preumont [93], Chao [21] and Sery and Napoly [118]. The reader is assumed to be familiar with basic terms of probability theory (see Papoulis and Pillai [80]), the continuous Fourier transform (see Pinsky [92]) and the discrete Fourier transform (see Oppenheim, Schafer and Buck [78]).

Before defining an STP mathematically, we want to introduce the concept with the help of an example. Due to the thermal motion of electrons in a resistor, a noise voltage $u(t)$ can be measured. This resistor voltage is an STP, since the outcome of the measurement is not predictable. Knowledge about the properties of this STP is an important

2. Modelling and simulation of ground motion effects

information, e.g., for the design of measurement devices.

It is necessary to quantify this intuitive introduction mathematically. The mathematical concept of an STP is very similar to the one of a *random variable* (RV) from probability theory. An RV maps every outcome of a random experiment $\omega \in \Omega$ onto $\mathbb{R}$ or a subset of it, where Ω is the set of all possible outcomes. To illustrate this concept, the throw of a die is one outcome ω of all possible throws Ω. The RV in this example is the assignment of the number of spots the dice shows, to the event. Contrary to an RV, an STP $X(t)$ maps the outcome of the random experiment ω not onto a real number, but onto a real-valued, time-dependent function $x(t)$. Additionally, for an arbitrary time $t = t_1$, $X(t_1)$ is an RV. Hence, an STP is an RV at every time step, and therefore an generalisation of an RV.

There are many STPs in nature that have a high relevance for science and engineering. Common examples, apart from electronics noise, are wind, ground motion (including earth quakes), temperature, and water waves exciting the motion of buildings, bridges, ships or oil platforms. A very important continuous STP, from a practical as well as a mathematical point of view, is the so-called *Wiener process* (also known as Brownian motion). Its behaviour can be easier explained by its time-discrete version, called *random walk*. A random walk is an STP that is created by adding a random number from a Gaussian distribution with zero mean to the current value $x[k]$ to create the value at the next time step $x[k+1]$. One of the two main ground motion models used in this work (ATL law, see Sec. 2.1.2) is such a random walk in time and space.

The Wiener process can be generalized to a class of STPs called *Lévi processes* also known as processes with *stationary independent increments* (SII). In SII processes, not only Gaussian random numbers with zero mean can be integrated to create an STP, but also differently distributed random processes (e.g. Poisson process). For completeness it should be mentioned that the Lévi process is a subclass of the more general *Markov processes*. Markov processes (of first-order) have the Markovian property, which is that the future value $x[k+1]$ is only dependent on the actual value $x[k]$. Using also values from earlier time steps does not add any new information.

2.1.1.1. Characterisation of one-dimensional stochastic processes

Using the fact that an STP is at each time step an RV, it can be fully described by the properties of the set of its RV. Therefore, a time-dependent first-order *probability density function* (PDF) can be defined as

$$f(\alpha, t) = \frac{\partial F(\alpha, t)}{\partial \alpha} \quad \text{with} \quad F(\alpha, t) = P(X(t) < \alpha), \tag{2.1}$$

where $F(\alpha, t)$ is called the *cumulative distribution function* (CDF) and $P(.)$ stands for the probability of a certain event. The first-order PDF $f(\alpha, t)$ characterises the STP at each time. To characterise also the relationship between different RV of the STP, higher order joint PDFs are used. As an example, the second-order PDF is defined by

$$f_X(\alpha_1, \alpha_2, t_1, t_2) = \frac{\partial F_X(\alpha_1, \alpha_2, t_1, t_2)}{\partial \alpha_1 \partial \alpha_2} \tag{2.2}$$

$$\text{with} \quad f_X(\alpha_1, \alpha_2, t_1, t_2) = P(X(t_1) < \alpha_1, X(t_2) < \alpha_2).$$

2.1. Ground motion

The higher the order of the used PDF, the more accurate the description of the stochastic process gets, since PDFs of lower order can be calculated from a higher order PDF. As a result of the Markovian property, every Markov process is fully characterised by its second order PDF.

Even though PDFs give a complete description of an STP, they are often complicate to work with. Therefore, it is convenient to extract *moments*. Most important are the first- and second-order moments, called *mean function* $\mu_X(t)$ and *autocorrelation function* $R_X(t_1, t_2)$. Using $\mathbb{E}\{.\}$ to symbolise the expectation value, these quantities are defined by

$$\mu_X(t) = \mathbb{E}\{X(t)\} = \int_{-\infty}^{+\infty} \alpha f_X(\alpha, t) \mathrm{d}\alpha \quad \text{and} \tag{2.3}$$

$$R_X(t_1, t_2) = \mathbb{E}\{X(t_1)X(t_2)\} = \int_{-\infty}^{+\infty}\int_{-\infty}^{+\infty} \alpha_1 \alpha_2 f_X(\alpha_1, \alpha_2, t_1, t_2) \mathrm{d}\alpha_1 \mathrm{d}\alpha_2. \tag{2.4}$$

An STP $X(t)$ is said to be *strict sense stationary* (SSS), if for arbitrary $n \in \mathbb{N}$ and $\tau \in \mathbb{R}$

$$f_X(\alpha_1, \alpha_2, \ldots, \alpha_n, t_1, t_2, \ldots, t_n) = f_X(\alpha_1, \alpha_2, \ldots, \alpha_n, t_1 + \tau, t_2 + \tau, \ldots, t_n + \tau). \tag{2.5}$$

From this property it follows, that the mean function is not changing over time $\mu_X(t) = \mu_X$, and that the autocorrelation function is only dependent on the time difference between the observed time points $R_X(t_1, t_2) = R_X(t_1 - t_2) = R_X(\tau)$, with $\tau = t_1 - t_2$. If only $\mu_X(t) = \mu_X$ and $R_X(t_1, t_2) = R_X(t_1 - t_2) = R_X(\tau)$ are true, the STP is called *wide sense stationary* (WSS). All ground motion models used in this text are based on the assumption that ground motion is a WSS STP.

Next, we want to present the main tool for the characterisation of STPs in this work, the so-called *power spectral density* (PSD). The PSD $P_X(\omega)$ describes how the average signal power of a WSS STP $X(t)$ is distributed over frequency. The PSD is defined by

$$P_X(\omega) = \lim_{T \to \infty} \frac{1}{T} \mathbb{E}\{|X(j\omega, T)|^2\}, \quad \text{with} \tag{2.6}$$

$$X(j\omega, T) \equiv \int_{-T/2}^{T/2} X(t)e^{-j\omega t} \mathrm{d}t, \tag{2.7}$$

where $X(j\omega, T)$ is called the truncated Fourier transfrom of $X(t)$. In Eq. (2.7), the integrant is the STP $X(t)$. The integration over an STP has to interpreted as an integral over each possible outcome $x_i(t)$ of $X(t)$. Hence also $X(j\omega)$ is an STP. The truncated Fourier transform has to be used instead of the Fourier transfrom, since an integration from $-\infty$ to $+\infty$ would result for most STP in an unbound integral.

The PSD can be used to evaluate the average power of an STP $X(t)$ coming from a certain frequency range. An often used quantity in this context is the integrated root mean square (IRMS) $\sigma_X(\omega_m)$. It is defined by the standard deviation of the amplitude of an STP, where only frequency components down to the angular frequency ω_m are considered. The IRMS can be calculated from the PSD by

$$\sigma_X(\omega_m) = \sqrt{2 \frac{1}{2\pi} \int_{\omega_m}^{\infty} P_X(\omega) \mathrm{d}\omega} \quad \text{with} \quad \omega_m \geq 0, \tag{2.8}$$

2. Modelling and simulation of ground motion effects

where the factor two in the nominator is introduced, since the integration is only carried out over positive angular frequencies and the PSD is an even function.

The *Wiener-Khinchine theorem* states that the PSD and the autocorrelation function of a wide sense stationary STP form a Fourier transform pair, which is

$$P_X(j\omega) = \int_{-\infty}^{+\infty} R_X(\tau)e^{-j\omega\tau}d\tau \quad \text{and} \quad R_X(\tau) = \frac{1}{2\pi}\int_{-\infty}^{+\infty} P_X(j\omega)e^{j\omega\tau}d\omega. \quad (2.9)$$

For practical applications, the characteristics of an STP have to be determined empirically. To determine the properties of an STP at a certain time t_1, several measurements would have to be conducted at the same time. This is obviously impossible. Therefore, all empirical characterisations rely on the *ergodicity* of the observed STP. Ergodicity is the property of an STP, that one outcome $x_i(t)$ of $X(t)$ contains all statistical information about the RV $X(t_j)$ for all $t_j \in \mathbb{R}$. A necessary condition for a process to be ergodic is that it is stationary. In most cases the ergodicity of a process can not be proven mathematically and the process is simply assumed to be ergodic (ergodic hypothesis).

For the following estimation of the PSD, we assume ergodicity of the underlying STP. Since measurements are usually available in sampled form x_k, the *discrete Fourier transform* (DFT) will be used. Thus, $P(\omega)$ can only be estimated at the positive angular frequencies $\omega_n = 2\pi n/T_0$, with $n = 1, 2, \ldots, N$, where N is the number of sampled points and T_0 the recording time. It can be shown (see Preumont [93]), that the Fourier-transform $X(j\omega_n)$ and the DFT $X[n]$ of a signal are related as

$$X(j\omega_n) = \frac{T_0}{N}X[n]. \quad (2.10)$$

In combination with the definition of the PSD in Eq. (2.6), Eq. (2.10) gives rise to estimate $P(\omega_n)$ as

$$P_{est1}(\omega_n) = \frac{T_0^2}{N^2 T_0}|X[n]|^2 = \frac{T_d}{N}|X[n]|^2 = \frac{1}{N^2\Delta f}|X[n]|^2, \quad (2.11)$$

where $T_d = T_0/N$ is the sampling time and $\Delta f = 1/T_0$ is the frequency resolution. The estimator $P_{est1}(\omega_n)$ is called *periodogram*. Unfortunately, $P_{est1}(\omega_n)$ turns out to be not a good estimator. As $T_0 \to \infty$, and therefore the number of recorded sampled $N \to \infty$, only the frequency resolution $1/T_0 \to 0$, but the variance of the individual spectral lines stays constant. Another procedure know as *Bartlett's method* has proven to be better suited to create an estimator for $P(\omega_n)$. The recorded time series $x[k]$ is cut into M equal parts $x^{(m)}[k]$, each with a recording time of T_0/M. The periodograms of all $x^{(m)}[k]$ are calculated and averaged to give the estimator

$$P_{est2}(\omega_i) = \frac{1}{\frac{T_0}{M}}\frac{\left(\frac{T_0}{M}\right)^2}{\left(\frac{N}{M}\right)^2}\frac{1}{M}\sum_{m=0}^{M-1}|X^{(m)}[i]|^2 = \frac{T_d}{N}\sum_{m=0}^{M-1}|X^{(m)}[i]|^2. \quad (2.12)$$

In case it is not necessary to calculate the PSD explicitly, but only the power of a certain signal has to be evaluated in the frequency domain, it is convenient to use *Parseval's theorem*

$$\sum_{i=1}^{N}|x[i]|^2 = \frac{1}{N}\sum_{n=1}^{N}|X[n]|^2. \quad (2.13)$$

2.1. Ground motion

If the left side of Eq. (2.13) is divided by N, the term corresponds to the average power of the signal $x[k]$. Therefore, $P[n] = \frac{1}{N^2}|X[n]|^2$ is a quantity, which can be used to calculate the power of a signal in the frequency domain by simply adding up its elements.

Finally, we want to state without proof, how an STP behaves when it is the input function for a dynamic system. When the stationary STP $X(t)$ is applied to a dynamic system with frequency response $H(j\omega)$, it can be shown that the output signal $Y(j\omega)$ of the system is also a stationary STP with the PSD

$$P_Y(\omega) = |H(j\omega)|^2 P_X(\omega). \tag{2.14}$$

2.1.1.2. Characterisation of multi-dimensional stochastic processes

The notion of an STP can be extended to *vector STP*, in which each element of a vector $\boldsymbol{X}(t)$ is an STP. Similarly to the scalar case a mean value vector

$$\boldsymbol{\mu_X}(t) = \mathbb{E}\left\{\boldsymbol{X}(t)\right\}, \tag{2.15}$$

and an autocorrelation matrix

$$\boldsymbol{R_X}(t_1, t_2) = \mathbb{E}\left\{\boldsymbol{X}(t_1)\boldsymbol{X}^H(t_2)\right\}, \tag{2.16}$$

can be defined, where ^{H} symbolises the conjugate transposed of a vector. With the help of the Wiener-Khinchine theorem (Eq. (2.9)) a PSD matrix can be calculated as

$$\boldsymbol{P_X}(\omega) = \int_{-\infty}^{+\infty} \boldsymbol{R_X}(\tau) e^{-j\omega\tau} d\tau. \tag{2.17}$$

While the diagonal elements correspond to the one-dimensional PSDs as in Eq. (2.9), the off-diagonal elements describe the mutual power between different STPs. It is convenient to normalise these *mutual power spectra*

$$N_{ij}(\omega) = \frac{P_{ij}(\omega)}{\sqrt{P_{ii}(\omega)P_{jj}(\omega)}}. \tag{2.18}$$

Of special importance is the real part of $N_{ij}(\omega)$, since it is a measure for the correlation of the STPs $X_i(t)$ and $X_j(t)$. If $N_{ij}(\omega) = 1$ both signals are fully correlated and move exactly the same way. If $N_{ij}(\omega) = 0$, $X_1(t)$ and $X_2(t)$ are fully uncorrelated and hence statistically independent.

Often STPs that are functions in time and space have to be described. An example would be ground motion, on the surface of the earth. In this case a description with a vector is not sufficient and a multi-dimensional function $X(\boldsymbol{s}_1, \boldsymbol{s}_2, t)$, called *random field*, has to be used, where $\boldsymbol{s}_1$ and $\boldsymbol{s}_2$ are two spatial points in some appropriate coordinates frame. Analogous to the vector STP case, a PSD function can be calculated for the multi-dimensional case as

$$P_X(\boldsymbol{s}_1, \boldsymbol{s}_2, \omega) = \int_{-\infty}^{+\infty} R_X(\boldsymbol{s}_1, \boldsymbol{s}_2, \tau) e^{-j\omega\tau} d\tau. \tag{2.19}$$

2. Modelling and simulation of ground motion effects

If the random field is spatially homogeneous, it depends only on the difference $s = s_1 - s_2$. In this case also a Fourier transform in the spatial domain can be performed

$$P_X(\mathbf{k}, \omega) = \int_{-\infty}^{+\infty} P_X(\mathbf{s}, \tau) e^{-j\mathbf{k}^T \mathbf{s}} d\mathbf{s}, \qquad (2.20)$$

where $\mathbf{k}$ is the vector of wave numbers. The term $P_X(\mathbf{k}, \omega)$ is the main tool in this work to describe ground motion and will be called the *two-dimensional ground motion PSD*. In this work, only one spatial coordinate will be considered and thus $\mathbf{k}$ will be a scalar and therefore $k = 2\pi/\lambda$, where λ is the wave length.

If $P_X(\mathbf{k}, \omega)$ is applied to a continuous structure, the PSD function of the output signal $P_Y(\mathbf{r}_1, \mathbf{r}_2, \omega)$, where $\mathbf{r}_1$ and $\mathbf{r}_2$ are two points in the output coordinates, can be calculated as

$$P_Y(\mathbf{r}_1, \mathbf{r}_2, \omega) = \int_k P_X(\mathbf{k}, \omega) G(\mathbf{r}_1, \mathbf{k}, \omega) G^H(\mathbf{r}_2, \mathbf{k}, \omega) d\mathbf{k} \qquad (2.21)$$

$$\text{with} \quad G(\mathbf{r}, \mathbf{k}, \omega) = \int_R H(\mathbf{r}, \mathbf{s}, \omega) e^{-j\mathbf{k}^T \mathbf{s}} d\mathbf{s}, \qquad (2.22)$$

where the *frequency response function* $H(\mathbf{r}, \mathbf{s}, \omega)$ is the amplitude of the output at position $\mathbf{r}$ due to a harmonic excitation at the input position $\mathbf{s}$. The expression $G(\mathbf{r}, \mathbf{k}, \omega)$ is called *sensitivity function*. The sensitivity functions calculated in this chapter will be time-independent and are only evaluated for some specific position $\mathbf{r}_{IP}$, which is the IP.

2.1.2. Ground motion models

The topic of ground motion is usually studied in the field of seismology. The phenomena interesting for seismology, such as e.g. tectonic movement and earthquakes, result in large ground motion amplitudes, but are nevertheless of little importance for the design of accelerators. Due to this reason, the accelerator community started in the 80's research activities to improve the understanding of ground motion in the regimes interesting for accelerators applications. The outcome of this activity is briefly summarised in the following, complimented by a discussion of some phenomena known from seismology (Fischer [39] and Aki and Richard [4]).

2.1.2.1. Seismic phenomena

Even though the surface of our earth seems to be very rigid, it behaves surprisingly elastic, if observed on a larger scale. The surface layer, called crust, is only about 7.5 to 35 km thick. It is not homogenous, but separated into tectonic plates, which float on the gooey to liquid layers below. Energy can propagate in the form of waves in this thin crust, but also through the inner layers of the earth. Also effects like ringing due to earthquakes have been observed.

The motion of the earth's crust can be excited by a variety of different sources. *Tectonic motion*, which is caused by the movement of tectonic plates against each other, can lead to a relative displacement of several centimetres per year. Since the site for a new accelerator can be chosen to be in an area with little or only moderate tectonic activity, these effects are considered to be not a problem for the accelerator operation.

2.1. Ground motion

Earthquakes are another seismic phenomenon, which originate, most of the time, from tectonic motion. When tectonic plates move against each other, the resulting stress can build up and is released at discrete time events. This results in shock-like ground motion of high amplitude. Since earthquakes are strong enough to disturb the accelerator operation are rare, they are only important for machine protection issues and civil engineering.

Also the varying *gravitational forces* of the sun and moon have effects on the earth. Not only the water level of the sea changes (tides), but also the crust of the earth is distorted. Even though these daily distortions can be in the centimetre range, they are of no concern for a particle accelerator. This is due to the fact that the distortions are smoothly distributed over a large area and create hardly any relative displacement throughout an accelerator site. Seismic effects that are important for the operation of an accelerator are *ground settlements* and *micro-seismic noise* and will be discussed in the following sections.

2.1.2.2. Diffusive ground motion and the ATL law

In seismology, ground motion is explained by the propagation of waves. For slow motion the wave length $\lambda = c_g/f$, where c_g is the wave propagation velocity, gets larger than the accelerator site and the earth itself. The use of waves is not appropriate anymore. In Baklakov et al. [6], measurements of the relative ground motion Δy are presented, which show that the variance $\sigma^2_{\Delta y}(T, L) = \mathbb{E}\left\{[y(s,t) - y(s+L, t+T)]^2\right\}$, where L is the spatial distance between two points and T is the time difference, scales as $\sigma^2_{\Delta y}(T, L) \propto TL$. This gives rise to the *ATL law*

$$\sigma^2_{\Delta y}(T, L) = ATL, \qquad (2.23)$$

where A is a site dependent constant. Note that the ATL law can only describe relative but no absolute motion. Since for an accelerator only relative motion is of importance, this does not result in any problems. Several measurements have been conducted, mainly at accelerator sites throughout the world (Shiltsev [121] and [120]). When large seismic components, e.g. tide effects, are removed from the measured spectra, all data sets follow the ATL law. The constant A varies strongly depending on the site from about 0.1×10^{-6} to $100 \times 10^{-6}\,\mu\text{m}^2/\text{m/s}$. The geometric mean of all measurements in Shiltsev [121] gives an A of $7 \times 10^{-6}\,\mu\text{m}^2/\text{m/s}$ and for CLIC the assumption of an A of $0.5 \times 10^{-6}\,\mu\text{m}^2/\text{m/s}$ was made. The assumption for CLIC is rather optimistic, but is consistent with the standard ground motion models A, B and B10, which will be explained in the next section.

From a mathematical point of view the ATL law is a two-dimensional random walk as described in Sec. 2.1.1. In general the ATL law represents a random field. Since over the area of interest this random field is assumed to be spatially homogenous, a two-dimensional PSD $P(k, \omega)$ as in Eq. (2.20) can be used to describe its properties. It can be shown (Sery and Napoly [118]) that the the ATL law can be represented by the PSD

$$P(k, \omega) = \frac{A}{k^2 \omega^2}. \qquad (2.24)$$

2. Modelling and simulation of ground motion effects

If only a limited number of spatial points s_i are observed, the random field can be simplified to a vector STP. If all M observed points are on a straight line as for a linac, ground motion data for computer simulations can be created with low numerically effort by

$$y[t_k, s_i] = y[t_{k-1}, s_i] + \sum_{m=1}^{i} \Delta[k, m] \quad \forall i : 1, \ldots, M \qquad (2.25)$$

with $y[t_0, s_i] = 0$,

where $\Delta[k, i]$ is calculated by

$$\Delta[k, i] = \sqrt{A[t_k - t_{k-1}]|s_i - s_{i-1}|} \mathcal{N}(0, 1), \qquad (2.26)$$

where $\mathcal{N}(0, 1)$ symbolises a Gaussian random number with zero mean and a variance of one. The initial values t_0 and s_0 determine the reference time and position. ATL motion generated by this method is shown in Fig. 2.2.

There are also attempts to physically explain the ATL law. Excitation sources as atmospheric pressure variations, wind and underground water flow are assumed to excite the ground. These excitations lead to a diffusive motion of the ground, which can be explained with a fractal model of the ground (Parkhomchuk and Shiltsev [81]). In this model the ground is separated into rigid blocks of different size, which model discontinuities of the properties of the ground.

It should also be mentioned that ground settlements in tunnels show a different behaviour than the ATL law. It was reported, e.g. in Sery and Raubenheimer [119], that the re-alignment of accelerators indicate a linear drift of the misalignment over time and therefore $\sigma^2_{\Delta y}(T, L) \propto T^2 L$. Even though these settling effects decrease over time, they appear still several years after the construction of the site. For time scales longer than one day, they become larger than the ATL motion and dominate the PSD. Hence, for studies of the accelerator behaviour up to one day, ATL motion is used, while longer periods are modelled by a similar procedure adapted to the ground settlement. Other seismic effects, as e.g. tides, should be eventually included in the used ground motion generator.

2.1.2.3. Wave-like motion

Especially ground motion components with higher frequencies can not completely be described by the ATL law. In this regime ground motion has mainly not a diffusive, but a wave-like character. These waves travelling through the crust of the earth contribute mainly to the absolute, but also to the relative motion.

The wave propagation in the earth can be studied with the theory of sound propagation in elastic media. Some important outcomes of this theory are collected here (please refer to Aki and Richard [4], Fischer [39] and Steinhagen, Redaelli and Wenninger [130] for more information). In a homogeneous, isotropic and elastic medium ground motion waves can propagate in two modes. The first one is a transversal polarised wave, often also called share wave or S-wave. The second mode of propagation is a longitudinal polarised wave, also called P-wave or pressure waves. On the surface of the earth the

2.1. Ground motion

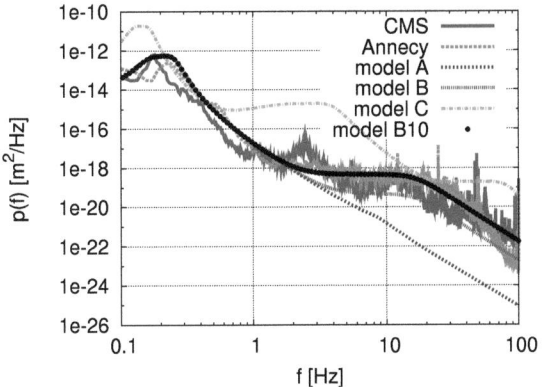

Figure 2.1.: Ground motion PSDs of measurements at different sites (CMS, Annecy) and ground motion models (A, B, B10, C). The micro-seismic peak at 0.1 to 0.3 Hz is clearly visible. The model B underestimates the cultural ($f > 2\,\text{Hz}$) noise of the measurements at CMS (Kuzmin [64]) and Annecy (Bolzon [12]). Therefore model B10 is created, which can be seen as a rather pessimistic assumption. (Picture courtesy of D. Schulte)

ground is not isotropic anymore, and the waves propagate in *Rayleigh* and *Love waves*, which are a superposition of P- and S-waves. For Rayleigh waves, the surface particles move on an ellipse in vertical and longitudinal direction with respect to the propagation direction of the wave. For Love waves, particles move on an ellipse in horizontal and longitudinal direction. Rayleigh waves are more important for accelerator applications. They travel with a velocity $v \approx \sqrt{E/(2\rho)}$, where ρ and E are the density and the Young's modulus E of the ground. The Young's module E is a measure of stiffness and relates the stress and the according elongation of a material.

Rayleigh and Love waves penetrate the ground approx. to a depth of the wavelength λ. Since E and ρ vary with the ground depth, the propagation velocity of waves depends also on their wavelength and hence their frequency. In Raubenheimer et al. [98] e.g., the empirical law

$$v(f) = 450 + 1900 e^{-f/2} \tag{2.27}$$

was found for the site of the Stanford Linear Accelerator Center (SLAC), which is consistent with the local ground properties.

Typical ground motion PSDs are shown in Fig. 2.1 (in this case only over frequency and not over the wave length). Without going into the details of the different curves ,which will be done later, it can be stated that all PSDs posse two characteristic features that exceed the ground motion contribution of the ATL law in the plotted frequency range. At a frequency of about 1/7 Hz the spectra show a significant peak that is called the *micro-seismic peak* or *7-sec hump*. It originates from swell waves in oceans, which couple to the coast. Swell waves are created by wind and storms over the oceans. Most

27

2. Modelling and simulation of ground motion effects

of the locally created waves (wind waves) have a relatively short wavelength and are dissipated quickly. Waves with longer λ are dissipated less and can travel over long distances. Observed at the coast, these swell waves show a strong directionality and narrow spectrum of λ. When the swell waves impinge on the shore only certain frequency components can couple to the ground, while others are reflected. The coupled energy is transported by ground waves far into the continent and contributes strongly to the short term ground motion.

The second important feature in Fig. 2.1 are humps at higher frequencies, which vary from site to site. These contributions originate from *cultural noise*, also called *technical noise*. This term is a collection for all man-made ground excitation, such as machinery in the accelerator tunnel (e.g. vacuum pumps), air ventilation and even traffic on the surface. The mentioned sources are local. The created waves spread and get attenuated by geometric spreading and dissipation. For the modelling of ground motion (two-dimensional PSD) the underlying STP is assumed to be spatial homogenous however. Thus, the complex geometric structure of cultural noise can not be fully represented with these models. For a ground motion sensitive accelerator like CLIC, all sources of cultural noise have to be identified and designed/chosen thoughtfully. Very important is also the design of the girder and the alignment system. If these elements are not stiff enough, ground motion could be amplified. For completeness it should be mentioned that there are also cultural noise sources not transmitted through the ground, such as cooling water flow, air flow from ventilation and sound waves. Studies of these sources are ongoing, but are not considered in this work.

2.1.2.4. Models based on the two-dimensional PSD

To model both, the diffusive motion of the ATL law and the wave-like components properly, the two-dimensional PSD $P(k,\omega)$ is used in the accelerator community. Sery and Napoly [118] describes a generic form of $P(k,\omega)$ that can be fitted to measurements by choosing certain open parameters. We will first review the generic model and will then discuss different parameter sets used in the literature.

The generic model for $P(k,\omega)$ has the form

$$P(\omega,k) = P_{ATL}(\omega,k) + \sum_{i=1}^{N} P_i(\omega,k), \qquad (2.28)$$

where $P_{ATL}(\omega,k)$ corresponds to a modified form of the ATL law and the N peaks $P_i(\omega,k)$ represent wave-like components. The basic ATL law in Eq. (2.24) had to be modified, since measurements did not show any diffusive motion for higher frequencies. Since the sensors have a limited precision all motion below the sensor precision was pessimistically assumed to be diffusion-like. This leads to the modified ATL law

$$P_{ATL}(\omega,k) = \frac{A}{\omega^2 k^2}\left[1 - \cos\left(\frac{Bk}{A\omega^2}\right)\right], \qquad (2.29)$$

where B is a constant defining the frequency where the $1/\omega^2$ behaviour of the ATL law changes to a $1/\omega^4$ behaviour for higher frequencies.

2.1. Ground motion

The wave-like components $P_i(\omega, k)$ represent peak in the spectrum and are realised by the expressions

$$P_i(\omega, k) = D_i(\omega) U_i(\omega, k) \quad \text{with} \tag{2.30}$$

$$D_i(w) = \frac{a_i}{1 + [d_i(\omega - \omega_i)/\omega_i]^4} \quad \text{and} \tag{2.31}$$

$$U_i(\omega, k) = \begin{cases} \frac{2}{\sqrt{k_i^2 - k^2}} & \text{if } |k| \leq k_i \\ 0 & \text{if } |k| > k_i \end{cases}. \tag{2.32}$$

The term $D_i(\omega)$ defines the form of the peak with respect to frequency, where a_i, d_i and ω_i are the height, the width and the position of the according peak. The term $U_i(\omega, k)$ determines the wave length composition of the peak, where k_i is the maximal occurring wave number given by $k_i = \omega/v_i$, where v_i is the wave propagation velocity. In current implementations of the model, v_i can also be chosen to be a function of ω as given in Eq. (2.27). Note that even though one individual wave of a certain wave length is assumed to travel over long distances, the complete signal does not have to have a high correlation. The correlation is determined by how many waves are propagating at the same time. A low v_i corresponds to a high k_i, which mean that the overall signal consists of waves with many different wave length and thus the correlation is low. On the other hand of v_i is high, the correlation of the corresponding signal is high. Hence, $U_i(\omega, k)$ defines the correlation properties of the peak. Low frequencies are well correlated (Juralev et al. [60]), while for higher frequencies >10 Hz the signals are nearly uncorrelated even over short distances. Measurements presented in Artoos et al. [5], also show the strong influence of separations in the concrete of the tunnel floor, which decrease the signal correlation strongly. This fact has to be considered at the construction of the tunnel.

Based on the generic model Eq. (2.28) several parameter sets were defined to model different ground motion conditions. In Sery and Napoly [118] four models are presented, which will be called model 1 to 4. Model 1, 2 and 3 represent tunnels with quiet conditions (no cultural noise) with different content of diffusive motion. These models originate from measurements performed in the tunnels of the Large Electron-Positron Collider (LEP), the Serpukhov accelerating-storage complex (UNK) and the Stanford Linear Accelerator Center (SLAC). Model 4 originates from measurements in the Hadron-Elektron-Ring-Anlage (HERA) tunnel and includes cultural noise, since the accelerator facilities were running during the measurements in this case.

The models 1 to 4 have been revised for the ILC-TRC report [1] and 3 new models A ("low"), B ("intermediate"), C ("high") have been created, which are a quasi-standard nowadays. These models will be used throughout this thesis. Their parameters can be found in Sery [115] and are collected with the parameters of model 1 to 4 in Tab. 2.1. Model A corresponds to a tunnel with very low ground motion. It originates from measurements at LEP and represents a deep tunnel in competent rock. Surface cultural noise has nearly no impact on the tunnel and in-tunnel technical noise is assumed to be very low. Model B originates from measurements in the Aurora mine near Fermilab. It stands for a shallow tunnel still on competent rock, in which cultural noise from the surface and the tunnel is at a medium level. In model C, the tunnel is shallow and built

2. Modelling and simulation of ground motion effects

		1	2	3	4	A	B	C
P_{ATL}:	A [μm^2/s/m]	10^{-4}	10^{-4}	10^{-6}	10^{-5}	10^{-7}	5×10^{-7}	10^{-5}
	B [μm^2/s^3]	10^{-3}	10^{-6}	10^{-6}	10^{-3}	5×10^{-7}	10^{-6}	5×10^{-6}
P_1:	f_1 [Hz]	0.14	0.14	0.14	0.14	10^{-3}	10^{-3}	0.14
	a_1 [μm^2/Hz]	10	10	10	10	10^3	10^3	10
	d_1 [1]	5	5	5	5	1	1	5
	v_1 [m/s]	10^3	10^3	10^3	10^3	3×10^3	2×10^3	10^3
P_2:	f_2 [Hz]	-	-	-	2.5	0.2	0.2	2.5
	a_2 [μm^2/Hz]	-	-	-	10^{-3}	0.35	0.35	10^{-3}
	d_2 [1]	-	-	-	1.5	3.5	3.5	1.5
	v_2 [m/s]	-	-	-	400	3×10^3	400	400
P_3:	f_3 [Hz]	-	-	-	50	5	4.5	50
	a_3 [μm^2/Hz]	-	-	-	10^{-7}	10^{-9}	2.5×10^{-8}	10^{-7}
	d_3 [1]	-	-	-	1.5	1.3	0.35	1.5
	v_3 [m/s]	-	-	-	400	3×10^3	400	400

Table 2.1.: Parameters of the most commonly used ground motion models. The parameters f_1, f_2 and f_3 correspond to ω_1, ω_2 and ω_3 in Eq. (2.31) and are related by $\omega_i = 2\pi f_i$. Model B10 is not listed in the table, since it is equivalent to model B, only that the third peak is amplified by a factor of ten, which leads to an a_3 of $10\times2.5\times10^{-8}$.

in a layer of sediment rock (measurements at HERA). The cultural noise corresponds to a strongly urbanised area (very high). In Fig. 2.1 recent measurements from the CMS experimental hall at CERN (Kuzmin [64]) and a setup in a laboratory in Annecy (France, Bolzon [12]) are shown. It can be seen that model B underestimates the cultural noise. Therefore, a new model B10 was introduced in which a corresponding cultural noise is amplified by a factor 10. This model B10 is a pessimistic assumption for an accelerator tunnel and will be mainly, but not exclusively, used in this thesis. For sake of completeness, we also want to mention other models in the literature. Model K originates from measurements at KEK and corresponds to very pessimistic assumptions, which are similar to the one for model C. Recently measurements have been conducted at CERN, which also include data for the ground motion in the horizontal plane (for according models see Collette et al. [25]).

The model in Eq. (2.28), in connection with the parameters in Tab. 2.1, can be used to estimate certain ground motion quantities. The one-dimensional PSD $P(\omega)$ can be calculated as (Sery and Napoly [118])

$$P(\omega) = \frac{1}{2\pi} \int_{-\infty}^{+\infty} P(\omega, k) \mathrm{d}k, \qquad (2.33)$$

to analyse only the frequency-dependence of ground motion. If the variance of the ground should be calculated, also an integration over ω has to be performed (absolute

2.1. Ground motion

motion)

$$\sigma^2 = \frac{1}{(2\pi)^2} \int_{-\infty}^{+\infty} \int_{-\infty}^{+\infty} P(\omega, k) \mathrm{d}k \mathrm{d}\omega. \tag{2.34}$$

For the analysis of the accelerator performance relative motion is more important than absolute. The variance of the relative motion after a time T can be derived as

$$\begin{aligned}\sigma_\Delta^2(T) &= \mathbb{E}\left\{[x(t+T) - x(t)]^2\right\} \\ &= \mathbb{E}\left\{x(t+T)^2\right\} - 2\mathbb{E}\left\{x(t+T)x(t)\right\} + \mathbb{E}\left\{x(t)^2\right\}.\end{aligned} \tag{2.35}$$

Since ground motion is assumed to be a stationary random field, the first and the last term are the same and can be substituted with Eq. (2.34). The middle term can be evaluated with the Wiener-Khinchine theorem in Eq. (2.9), which results in

$$\sigma_\Delta^2(T) = \frac{1}{(2\pi)^2} \int_{-\infty}^{+\infty} \int_{-\infty}^{+\infty} P(\omega, k) 2\left[1 - e^{j\omega T}\right] \mathrm{d}k \mathrm{d}\omega. \tag{2.36}$$

Taking into account that the PSD is an even function and applying Euler's formula, this expression can be simplified to

$$\sigma_\Delta^2(T) = \frac{1}{(2\pi)^2} \int_{-\infty}^{+\infty} \int_{-\infty}^{+\infty} P(\omega, k) 2\left[1 - \cos(\omega T)\right] \mathrm{d}k \mathrm{d}\omega. \tag{2.37}$$

The factor $2[1-\cos(\omega T)]$ varies from 0 to 4 and describes the change of the signal power, if not the sine signal itself is considered, but the differential motion of the sine wave after a time difference T. As an example, for $0 \leq \omega T \ll \pi/2$ the wave moves only little and the power of the PSD is demagnified. If on the other hand $\omega T = \pi(2n-1)$, with $n \in \mathbb{Z}$, the power of the differential signal is a factor 4 larger than the original signal.

In case also a spatial distance L is taken into account, the differential motion (assuming perfect initial alignment) can be calculated with a similar approach as in Eq. (2.35) as

$$\sigma_\Delta^2(T, L) = \frac{1}{(2\pi)^2} \int_{-\infty}^{+\infty} \int_{-\infty}^{+\infty} P(\omega, k) 2\left[1 - \cos(\omega T)\right] 2\left[1 - \cos(kL)\right] \mathrm{d}k \mathrm{d}\omega. \tag{2.38}$$

For simulations it is necessary to create samples in the time and spatial domain out of the two-dimensional PSD. A standard method for this task is presented in Preumont [93], which uses the inverse DFT. The following method is equivalent to the one in Preumont [93], but omits the inverse DFT in order to give more physical insights. The ground motion generator used for the integrated simulations in Sec. 2.4 uses the described method. If a signal $x(s,t)$ with variance σ_x^2 and a PSD $P_X(\omega, k)$ should be created, it is useful to partition the power of the PSD into discrete portions

$$\begin{aligned}\sigma_x^2 &= \frac{1}{(2\pi)^2} \int_{-\infty}^{+\infty} \int_{-\infty}^{+\infty} P(\omega, k) \mathrm{d}\omega \mathrm{d}k \\ &\approx \frac{1}{(2\pi)^2} \sum_{i=0}^{N_\omega} \sum_{j=1}^{N_k} \left(4 \int_{\omega_i}^{\omega_{i+1}} \int_{k_j}^{k_{j+1}} P(\omega, k) \mathrm{d}\omega\right) \mathrm{d}k = \sum_{i=0}^{N_\omega} \sum_{j=1}^{N_k} a_{ij}^2\end{aligned} \tag{2.39}$$

with $\quad a_{ij} = \frac{1}{\pi} \sqrt{\int_{\omega_i}^{\omega_{i+1}} \int_{k_j}^{k_{j+1}} P(\omega, k) \mathrm{d}\omega \mathrm{d}k}. \tag{2.40}$

2. Modelling and simulation of ground motion effects

The ω_i and k_j are chosen to be positive values, which cover all significant components of $P(\omega, k)$. Since $P(\omega, k)$ is a symmetric function and the ω_i and k_j only run over positive values, a factor 4 has to be added to account for negative k and ω. The a_{ij}^2 corresponds to signals with a given power content and a defined frequency and wave number $\hat{\omega}_i = (\omega_{i+1} + \omega_i)/2$ and $\hat{k}_j = (k_{i+1} + k_j)/2$. This corresponds to a travelling wave of the form

$$x_{ij}(s,t) = \sqrt{2} a_{ij} \sin(\hat{k}_j s + \hat{\omega}_i t + \phi_{ij}), \tag{2.41}$$

where the phases ϕ_{ij} are taken from an uniformly distributed random generator in the interval $[0, 2\pi)$. The overall signal with the correct PSD can be created by

$$x(s,t) = \sum_i \sum_j x_{ij}(s,t). \tag{2.42}$$

The resulting STP in Eq. (2.42) is a Gaussian RV at any point in time and space, due to the following argument. Because of the properties of the chosen ϕ_{ij}, for any time t_0 and position s_0, the $x_{ij}(s_0, t_0)$ are independent random numbers. By the central limit theorem, the final signal is a Gaussian distributed RV, since the sum of a large number of independent RV is Gaussian distributed. In Fig. 2.2 random signals generated with this procedure are shown. The current version of the ground motion generator in PLACET uses some modifications to reduce the computational complexity and to implement filter functions. These details will be given in Sec. 2.4.

2.1.2.5. Effect of the stabilisation system

The effective motion of a stabilised quadrupole is altered by the stabilsation system introduced in Sec. 1.3.2.4. The action of this system can be modelled by its frequency response $S_{ST}\left(e^{j\omega T_{d,ST}}\right)$ (see Fig. 1.8 (right)), where $e^{j\omega T_{d,ST}}$ symbolises the discrete-time nature of the stabilisation system (see App. C.2) and $T_{d,ST}$ is its sampling time.

To calculate for example the variance of the absolute motion of a stabilised quadrupole, Eq. (2.34) has to be modified to

$$\sigma^2 = \frac{1}{(2\pi)^2} \int_{-\infty}^{+\infty} \int_{-\infty}^{+\infty} P(\omega, k) \left| S_{ST}\left(e^{j\omega T_{d,ST}}\right) \right|^2 \mathrm{d}k \mathrm{d}\omega. \tag{2.43}$$

2.2. Luminosity and beam orbit impact of ground motion

In this section, the ground motion models introduced in Sec. 2.1 will be used to analyse the effect of ground motion on certain beam parameters. There are two types of models for beam parameter variations due to ground motion used in the literature: analytic and PSD-based models. Both types are reviewed in Sec. 2.2.1 before they are adapted and extended for the needs of the controller design for CLIC in Sec. 2.2.2 and Sec. 2.2.3. The scope and the limitations of the developed models will be discussed.

2.2. Luminosity and beam orbit impact of ground motion

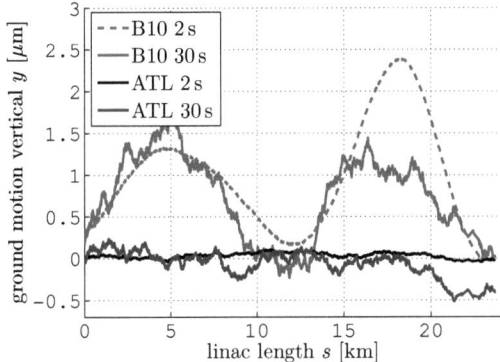

Figure 2.2.: Vertical ground motion along the main linac and BDS of CLIC generated by a ground motion generator. The evolution of ground motion generated according to the ATL law (blue and black) and model B10 (red) are compared. The motion from model B10 reaches already after 2 s large amplitudes due to the microseismic peak. After longer time periods (30 s) the B10 motion gets only insignificantly larger but less smooth, due to the appearing ATL components. Also, motions created according to the ATL model are plotted for 2 and 30 s.

2.2.1. Basics

Two basic types of models to estimate the effect of ground motion on beam parameters can be found in the literature. The first type delivers simple, analytic formulas to produce estimates of the effects to be expected. To be able to derive these formulas the accelerator component motion (quadrupoles, accelerating structures ...) has to be assumed to be of a simple form, e.g. independent white noise or ATL excitation.

To be able to include more complex ground motion behaviour, a second class of models is available. It is based on the two-dimensional ground motion PSD and more computational effort is needed to evaluate them. We will in the following introduce these two types of models by giving representative examples. References to the literature will be given to ease further studies.

2.2.1.1. Analytical models

We derive in the following one exemplary model with the intention to familiarise the reader with the basic techniques used for the deduction of simplified analytical models. The derived model describes the beam jitter at the end of the main linac of CLIC, due to uncorrelated quadrupoles misalignments. After this derivation, the literature about analytic models will be reviewed. The reader is assumed to be familiar with basics terms of beam physics. For an introduction please refer to Wille [136], Wiedemann [135] or Holzer [2].

2. Modelling and simulation of ground motion effects

The creation of the model of beam jitter is split up in two parts. First, a general expression (Eq. (2.47)) will be derived by following the work of Kubo [63]. Then this expression will be specialised for the CLIC main linac in Eq. (2.51) using approximations from Schulte [112].

The beam position $y_f[i]$ and the beam angle $y'_f[i]$ at the end a beam line (symbolised by f), due to a dipole kick $\Theta[i]$ at a certain position indexed with i can be written as

$$y_f[i] = \sqrt{\beta_f \beta[i]} \sin \phi[i] \sqrt{E[i]/E_f} \Theta[i] \tag{2.44}$$

$$y'_f[i] = \sqrt{\beta_f \beta[i]} \left(\cos \phi[i] - \alpha[i] \sin \phi[i] \right) \sqrt{E[i]/E_f} \Theta[i]. \tag{2.45}$$

The terms β and α are the usual twiss parameters, $\phi[i]$ is the phase advance from position s_i to s_f and E is the beam energy. The expressions Eqs. (2.44) and (2.45) can be derived by using normalised coordinates (see Schulte [112]). The adiabatic damping due to the acceleration is accounted for by the factor $\sqrt{E(i)/E_f}$. The kick $\Theta[i]$ of a quadrupole misaligned by $\delta[i]$ is $\Theta[i] = -k[i]\delta[i]$, where $k[i]$ is the integrated strength of a magnet.

If more than one quadrupole is considered, the overall beam position y_f and beam angle y'_f is calculated as $y_f = \sum_i y_f[i]$ and $y'_f = \sum_i y'_f[i]$. A quantity covering both y_f and y'_f is the action of the beam oscillations J_f, which is defined by

$$J_f = \left(\gamma_f y_f^2 + 2\alpha_f y_f y'_f + \beta_f y_f'^2 \right)/2, \tag{2.46}$$

where γ_f is the third twiss parameter at the position s_f. By assuming that the kicks $\Theta[i]$ are uncorrelated and using Eqs. (2.44) and (2.45) in Eq. (2.46), the expectation value of J_f can be evaluated as

$$\mathbb{E}\{J_f\} = \frac{\sigma_\delta^2}{2E_f} \sum_i \beta[i] k[i]^2 E[i], \tag{2.47}$$

where $\sigma_\delta^2 = \mathbb{E}\{\delta^2\}$ is the variance of the quadrupole displacement. The general expression in Eq. (2.47) can be specialised for the case of CLIC. For this reason, it is useful to apply the technique of *smooth approximation*, which converts the sum over some arbitrary accelerator parameters $T[i]$, which are a function of the position index i, in an integral over the energy by

$$\sum_i T[i] \quad \rightarrow \quad \int_{E_0}^{E_f} \frac{T(E)}{\Delta E(E)} dE, \tag{2.48}$$

where $\Delta E(E)$ is the energy gain between two quadrupoles. With this approximation Eq. (2.47) becomes

$$\mathbb{E}\{J_f\} = \frac{\sigma_\delta^2}{2E_f} \int_{E_0}^{E_f} \frac{\beta(E) k(E)^2 E}{\Delta E(E)} dE. \tag{2.49}$$

Using the approximations (taken from Schulte [112])

$$\beta(E) = \frac{4}{k(E)^2 L(E)} \frac{1}{\sqrt{\frac{4}{k(E)^2 L(E)^2} - 1}}, \qquad k(E) = k_0 \sqrt{\frac{E_0}{E}},$$

$$\Delta E(E) = L(E) \eta_{fill} G e \quad \text{and} \qquad L(E) = L_0 \sqrt{\frac{E}{E_0}}, \tag{2.50}$$

where $L(E)$ is the average distance between two quadrupoles as a function of the energy, η_{fill} is the fill factor of the accelerating structures, G is the acceleration gradient of the structures and e is the charge of an electron. Substituting these terms into Eq. (2.49) and solving the integral gives the expression

$$\mathbb{E}\{J_f\} = \frac{2E_0(E_f - E_0)}{L_0^2 E_f \eta_{fill} Ge} \frac{1}{\sqrt{\frac{4}{k_0^2 L_0^2} - 1}} \sigma_\delta^2. \tag{2.51}$$

This expression of the average action can be used to calculate the variance of the beam offset $\mathbb{E}\{y_f^2\}$ normalised to the nominal beam size squared σ_y^2 at the end of the linac by

$$\frac{\mathbb{E}\{y_f^2\}}{\sigma_y^2} = \frac{\mathbb{E}\{J_f\}\beta_f}{\epsilon_N \beta_f/\gamma} = \frac{\mathbb{E}\{J_f\}\gamma}{\epsilon_N}, \tag{2.52}$$

where ϵ_N is the normalised beam emittance and γ is the relativistic factor. Combining Eqs. (2.51) and (2.52) and evaluating the result for the parameter of the main linac of CLIC for the vertical direction gives the final estimate

$$\frac{\mathbb{E}\{y_f^2\}}{\sigma_y^2} \approx 2.5 \times 10^{-3} \sigma_\delta^2, \tag{2.53}$$

where σ_δ has to be given in nm in this equation. Using Eq. (2.53), it can be deducted that if the average, normalised beam jitter should be kept below 1%, the quadrupole motion variance has to be smaller than 2 nm.

The main reference for analytic models of ground motion effects on beam parameters is Raubenheimer [97], which gives estimates for beam jitter, due to uncorrelated and ATL displacement of quadrupoles. Also an estimate for the emittance growth due to ATL motion is given. While Raubenheimer only states the final results, Kubo [63] gives more insights and adds an estimate for the emittance growth due to uncorrelated ground motion. Not only the effect of quadrupoles is studied but also the effect of accelerating cavities (see again Raubenheimer [97] and Kubo [63]). In an earlier work of Raubenheimer [96], estimates for the effect of a wave-like displacement of the quadrupoles are derived. Effects due to wave-like excitation for rings have been collected in Chao and Tigner [21].

2.2.1.2. Models based on the ground motion PSD

The analytic models discussed in the last section are not capable to include more complex ground motion models which are usually given by the two-dimensional PSD. Also effects that are not properly described by the usual linear beam optics—as wake fields, filamentation, and beam-beam effects—can not be fully covered, even thought they show linear or quadratic behaviour over large ranges of accelerator component displacement.

Basic formalism To include more complex ground motion models and beam physics effects a different type of model is introduced in Sery and Napoly [118]. In this model the

2. Modelling and simulation of ground motion effects

variances $\sigma_z^2 = \mathbb{E}\left\{z^2\right\}$ including feedback effects, where z is a generic beam parameter, can be calculated by extending Eq. (2.43) to

$$\sigma_z^2 = \frac{1}{(2\pi)^2} \int_{-\infty}^{+\infty} \int_{-\infty}^{+\infty} P(\omega,k) \left|S_{ST}(e^{j\omega T_{d,ST}})\right|^2 \left|\hat{S}(e^{j\omega T_d})\right|^2 G_z^2(k) \mathrm{d}\omega \mathrm{d}k. \qquad (2.54)$$

In this expression $\hat{S}(e^{j\omega T_d})$ and $S_{ST}(e^{j\omega T_{d,ST}})$ are the frequency responses of the L-FB and the quadrupole stabilisation to ground motion (sensitivity transfer function), and T_d is the sampling time, which is given by the beam repetition rate of 20 ms. Since the feedback algorithm is a discrete-time system, the frequency response is written in terms of $e^{j\omega T_d}$ to symbolise its periodic structure (see App. C.2). The function $G_z(k)^2$ describes the effect due to a misalignment of the accelerator with a harmonic wave onto the observed beam parameter z. Hence, $G_z(k)$ is a sensitivity function as defined in Eq. (2.22). Note that the feedback system acts, in this model, on the ground motion itself, which is a simplification to reality, where the L-FB acts mainly on the beam oscillations. Note further that in Eq. (2.54) the implicit assumption is made that the feedback action is only dependent on ω and not on k. This is not the case for the L-FB developed in Chap. 3. The consequences will be discussed in Sec. 2.2.4.

Since Eq. (2.54) differs in two ways from its original form in Sery and Napoly [118] (Eq. (60)), we want to comment on the variations. The first is the nomenclature: $F_{SERY}(\omega) = |\hat{S}(e^{j\omega T_d})|^2$ and $G_{SERY}(k) = G_z^2(k)^2$. This difference originates only from different definitions and does not change the final results for σ_z^2. We introduced a different notation to be consistent with the definition of the sensitivity function not considered by Sery. Secondly, the expression for σ_z^2 in Sery is by a factor 2 larger than in Eq. (2.54). This is due to the fact that Sery's expression corresponds to the variance of the beam parameter between two time steps t and $t+T$, while our expression corresponds to the variance of the absolute beam parameter at an arbitrary time t. Since the beam parameters at t and $t+T$ are uncorrelated random variables, the differential variance is 2 times larger than the absolute variance. Also the fact that for the simulations in this thesis a perfectly aligned accelerator is used at t does not change this statement, since alignment is only a static change of the accelerator component positions, leaving the underlying ground motion unchanged. We argue that the absolute beam parameter variance is more relevant than the relative and use therefore Eq. (2.54) throughout this thesis.

As an example, Eq. (2.54) will be used to evaluate the quadrupole motion $q(t)$ due to ground motion. In this case the integration over ω is not carried out in order to be able to resolve the frequency dependency, which leads to the PSD

$$Q_c(\omega) = \frac{1}{2\pi} \int_{-\infty}^{+\infty} P(\omega,k) \left|S_{ST}(e^{j\omega T_{d,ST}})\right|^2 \left|\hat{S}(e^{j\omega T_d})\right|^2 G_q^2(k) \mathrm{d}k. \qquad (2.55)$$

The expression for $Q_c(\omega)$ has been evaluated assuming the quadrupole stabilisation system V1 as described in Sec. 1.3.2.4 and the L-FB transfer function as will be given in Eq. (3.52) with a gain factor f_i of 0.35. The sensitivity function $G_q(k)$ for the quadrupole motion is equal to one. If the continuous motion $q(t)$ is sampled with the beam repetition rate to $q[k]$, the discrete-time spectrum $Q_d(\omega)$ of $q[k]$ is created by the folding of $Q_c(\omega)$ due to the aliasing effect. Both spectra $Q_c(\omega)$ and $Q_d(\omega)$ are depicted in Fig. 2.3.

2.2. Luminosity and beam orbit impact of ground motion

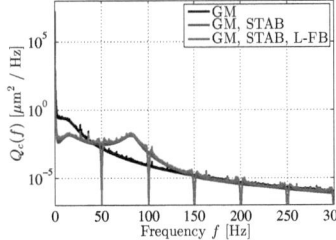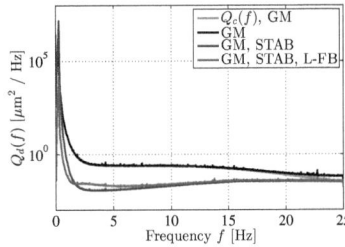

Figure 2.3.: PSDs $Q_c(f)$ (left) and $Q_d(f)$ (right) of the continuous-time and sampled quadrupole motion $q(t)$ and $q[k]$, if only ground motion excitation (GM) is considered or also the stabilisation system V1 (STAB) and the orbit feedback (L-FB) are used. The slight ripple on the curves originates from the limited numerical accuracy. (left) While the stabilisation system acts like a continuous-time filter on the original spectrum, the orbit feedback shows a periodic behaviour, which is typical for discrete-time filters. (right) The maximal resolvable frequency of 25 Hz of the discrete-time spectrum is determined by the sampling rate of 20 ms. While the stabilisation system reduces the quadrupole motion in the wide frequency range from about 1 to 25 Hz, the orbit feedback only demagnifies the motion for frequencies below 2 Hz. For comparison also the first part of the continuous-time PSD $Y_c(f)$ with only ground motion excitation is plotted. The aliasing effect only affects frequencies around 25 Hz slightly.

Determination of $G_z(k)$ The term $G_z(k)$ can be calculated in two ways: determination by simulations (accurate, comfortable but slow) or via analytical derivation (often not possible to cover all effects accurately, but fast). Both methods will be discussed in the following before they will be used to determine sensitivity functions for the beam oscillations in the BPMs (by simulations in Sec. 2.2.2), the luminosity loss (analytically in Sec. 2.2.3) and the beam-beam offset (analytically in Sec. 2.3.1). The resulting models are essential for the feedback design in Chap. 3.

First the numerical calculation with the help of simulation tools is covered. If a travelling wave with wave number k

$$y(s,t) = A(k)\sin(sk - \omega t - \phi_0) \qquad (2.56)$$

travels along the linac in positive direction, then also an observed beam parameter $z(t,k) = B(k)\sin(\omega t + \phi_z)$ is a harmonic function in time, where $B(k)$ is the amplitude and ϕ_z the phase of the oscillation. This statement follows from Eq. (2.54). The amplitude $B(k)$ is only dependent on k, since the accelerator system is a static system. The sensitivity function $G_z(k)$ is now defined as the amplitude ratio of the ground motion and the parameter wave $G_z(k) = B(k)/A(k)$. This motivates the following procedure to determine $G_z(k)$. The accelerator components of interest are misaligned with a harmonic function of wave number k and amplitude $A(k)$ as in Eq. (2.56) with $\phi_0 = 0$. The beam is tracked through the misaligned accelerator and the parameter z is recorded. Then the phase ϕ_0 is changed by steps of the size π/N, $N \in \mathbb{N}$, and the procedure is

2. Modelling and simulation of ground motion effects

repeated. The largest found value of z corresponds to $B(k)$ and $G_z(k)$ can be calculated by $G_z(k) = B(k)/A(k)$. The same procedure is repeated for every k. The parameter N has to be chosen large enough, such that the observed harmonic function of z is resolved finely enough. The excitation amplitude $A(k)$ has to be chosen large enough, such that the value of z is large compared to the simulation noise, but small enough that z stays in the linear regime. Note that an appropriate value for $A(k)$ can vary strongly for different wave lengths, and has to be found empirically (brute force simulations for many different values of $A(k)$ over all k).

Since an iteration over many different phases $\pi n/N$, $n = 0, \ldots, N-1$ is computational expensive it is useful to apply the following strategy, which reduces the number of iterations from N to 2. The traveling wave in Eq. (2.56) can be split up into two standing waves by using the trigonometric identity $\sin(a - b) = \cos(b)\sin(a) - \sin(b)\cos(a)$ as

$$y(s,t) = A(k)\sin(sk - \omega t - \phi_0)$$
$$= A(k)\cos(\omega t + \phi_0)\sin(sk) - A(k)\sin(\omega t + \phi_0)\cos(sk). \qquad (2.57)$$

Since the accelerator system is assumed to be linear and static (no phase shift in time) the components $\sin(sk)$ and $\cos(sk)$ can be treated independently. The observed parameter z will have the form

$$z(t,k) = A(k)A_s(k)\cos(\omega t + \phi_0) - A(k)A_c(k)\sin(\omega t + \phi_0)$$
$$= A(k)G_z(k)\sin(\omega t + \phi_z) \qquad (2.58)$$

with $\quad G_z(k) = \sqrt{A_s(k)^2 + A_c(k)^2} \quad$ and $\quad \phi_z = \phi_0 - \arctan\left(-\frac{A_c(k)}{A_s(k)}\right), \qquad (2.59)$

where $A_s(k)$ and $A_c(k)$ are the amplitudes of the parameter z, due to the ground motion excitation with a spatial sine or cosine wave of unit amplitude. Hence, the simulation only has to be performed for one sine and one cosine excitation.

The second method to determine $G_z(k)$ is by analytical derivation. The approach is similar to the analytically derived simple models in Sec. 2.2.1.1, but differs in two ways. The motion of the accelerator components, and more importantly the differential motion between components, is expressed with the help of the two-dimensional PSD and not with simple quantities as the variance. Additionally, the effect of a misaligned accelerator element on the beam parameter of interest is not expressed by the twiss parameters of the lattice. Instead, the change of the parameter z, due to a displacement $\delta[i]$ is modelled linearly by $z = a_z[i]\delta[i]$, where the parameter $a_z[i]$ is defined as $a_z[i] = \mathrm{d}z/\mathrm{d}\delta[i]$ and can be determined by simulations. Due to the determination of $a_z[i]$ by simulation, effects which are not covered by the description with twiss parameters can be included in the model, such as wake field and filamentation. In the next section, it will be shown $G_z(k)$ can be derived from the two-dimensional PSD and the parameters $a_z[i]$.

2.2.2. Sensitivity function for beam oscillations

In the following a sensitivity function $G_b(k,i)$ is derived analytically. This function describes the beam offset $b[i]$ in the ith BPM, due to a misalignment of the quadrupoles with a travelling wave with wave number k. The sensitivity function $G_b(k,i)$ will be an

2.2. Luminosity and beam orbit impact of ground motion

essential ingredient for the design of the L-FB (see also Sec. 2.2.4). The derivation is similar to the one of Sery and Napoly [118], where also expressions for the beam-beam offset at the IP and beam size growth due to dispersion are given.

Using the index i for BPMs and j for quadrupoles, $y[j]$ for the quadrupole misalignment, y_B for the initial beam offset, $\tilde{y}[i]$ for the BPM displacement, and N_i as the index of the last quadrupole influencing the ith BPM, then the beam offset $b[i]$ in the ith BPM is given by

$$b[i] = \tilde{r}_i y_B + r_{i,1}y[1] + r_{i,2}y[2] + \cdots + r_{i,N_i}y[N_i] - \tilde{y}[i], \qquad (2.60)$$

where the $r_{i,j}$, defined by $r_{i,j} = \mathrm{d}b[i]/\mathrm{d}x[j]$, correspond to the $a[i]$ in the general description in Sec. 2.2.1.2. The $r_{i,j}$ form the elements of the orbit response matrix $\boldsymbol{R}$, described in more detail in Sec. 3.2.1. The term $-y[i]$ in Eq. (2.60) takes into account that the BPM reading is also influenced by the motion of the BPM itself, and not only by the beam motion. The term $\tilde{r}[i] = \mathrm{d}b[i]/\mathrm{d}y_B$ accounts for the beam motion due to an initial beam offset. Note that $\tilde{r}[i]$ is not independent of the $r_{i,j}$. This becomes obvious if the whole, perfectly aligned beam line is moved by a constant value $y[i] = \hat{y}$ and also the initial beam offset $y_B = \hat{y}$. Since the complete system is moved, the BPM reading has still to be zero. Applying these values to Eq. (2.60) gives

$$0 = (\tilde{r}_i + r_{i,1} + \cdots + r_{i,N_i} - 1)\hat{y}$$

$$\rightarrow \quad \tilde{r}_i = 1 - \sum_{j=1}^{N_i} r_{i,j}. \qquad (2.61)$$

The beam is assumed to enter in the centre of the first element of the displaced beam line. Therefore, $y_B = y_0$, where y_0 is the ground motion displacement at the start of the accelerator. To shorten notation Eq. (2.60) is written as

$$b[i] = \sum_{j=0}^{N_i+1} r_{i,j}y[j] \quad \text{with} \qquad (2.62)$$

$$r_{i,0} = \tilde{r}_i, \qquad\qquad y[0] = y_B = y_0,$$
$$r_{i,N_i+1} = -1 \quad \text{and} \qquad y[N_i+1] = \tilde{y}[i].$$

The variance of the BPM reading $\sigma_b^2[i]$ can now be written as

$$\sigma_b^2[i] = \mathbb{E}\left\{b^2[i]\right\} = \mathbb{E}\left\{\left(\sum_{j=0}^{N_i+1} r_{i,j}y[j]\right)\left(\sum_{j=0}^{N_i+1} r_{i,j}y[j]\right)\right\}$$
$$= \mathbb{E}\left\{\sum_{m=0}^{N_i+1}\sum_{n=0}^{N_i+1} r_{i,m}r_{i,n}y[m]y[n]\right\} = \sum_{m=0}^{N_i+1}\sum_{n=0}^{N_i+1} r_{i,m}r_{i,n}\mathbb{E}\left\{y[m]y[n]\right\}. \qquad (2.63)$$

By using the Wiener-Khinchine theorem in Eq. (2.9) for the spatial dimension and considering the fact that the two-dimensional ground motion PSD $P(\omega, k)$ and the absolute value of the frequency response of the the quadrupole stabilisation $S_{ST}(e^{j\omega T_{d,ST}})$ are

2. Modelling and simulation of ground motion effects

even functions, the correlation $\mathbb{E}\{y[m]y[n]\}$ can be written as

$$\mathbb{E}\{y[m]y[n]\} = \frac{1}{(2\pi)^2} \iint_{-\infty}^{+\infty} P(\omega,k) \left|S_{ST}(e^{j\omega T_{d,ST}})\right|^2 e^{ikL_{m,n}} d\omega dk$$

$$= \frac{1}{(2\pi)^2} \iint_{-\infty}^{+\infty} P(\omega,k) \left|S_{ST}(e^{j\omega T_{d,ST}})\right|^2 \left[\cos(kL_{m,n}) + j\sin(kL_{m,n})\right] d\omega dk$$

$$= \frac{1}{(2\pi)^2} \iint_{-\infty}^{+\infty} P(\omega,k) \left|S_{ST}(e^{j\omega T_{d,ST}})\right|^2 \cos(kL_{m,n}) d\omega dk, \quad (2.64)$$

where $L_{m,n}$ is the distance between the quadrupoles m and n. Using Eq. (2.64) in Eq. (2.63) and exchanging summation and integration (possible since summation is finite) results in

$$\sigma_b^2(i) = \frac{1}{2\pi} \int_{-\infty}^{+\infty} B(\omega,i) d\omega \quad \text{with} \quad (2.65)$$

$$B(\omega,i) = \frac{1}{2\pi} \int_{-\infty}^{+\infty} P(\omega,k) \left|S_{ST}(e^{j\omega T_{d,ST}})\right|^2 G_b(k,i)^2 dk \quad \text{and} \quad (2.66)$$

$$G_b(k,i) = \sqrt{\sum_{m=0}^{N_i+1} \sum_{n=0}^{N_i+1} r_{i,m} r_{i,n} \cos(kL_{m,n})}, \quad (2.67)$$

where $B(\omega,i)$ is the PSD of the beam offsets in the ith BPM. On the first view the expression for $G_b(k,i)$ seems to be uncomfortable to evaluate. An easier and more intuitive representation can be achieved by using

$$\cos(L_{m,n}k) = \cos((s_n - s_m)k) = \cos(s_m k)\cos(s_n k) + \sin(s_m k)\sin(s_n k), \quad (2.68)$$

where s_m and s_n are the positions of the quadrupoles m and n. Using Eq. (2.68) in Eq. (2.67) gives

$$G_b(k,i)^2 = \sum_{m=0}^{N_i+1} \sum_{n=0}^{N_i+1} r_{i,m} r_{i,n} \cos(s_m k)\cos(s_n k)$$

$$+ \sum_{m=0}^{N_i+1} \sum_{n=0}^{N_i+1} r_{i,m} r_{i,n} \sin(s_m k)\sin(s_n k)$$

$$= \left[\boldsymbol{r}[i]^T \boldsymbol{c}(k,i)\right]^2 + \left[\boldsymbol{r}[i]^T \boldsymbol{s}(k,i)\right]^2 \quad (2.69)$$

with

$$\boldsymbol{r}[i] = \begin{bmatrix} r_{i,0} \\ r_{i,1} \\ \vdots \\ r_{i,N_i+1} \end{bmatrix}, \quad \boldsymbol{c}(k,i) = \begin{bmatrix} \cos(s_0 k) \\ \cos(s_1 k) \\ \vdots \\ \cos(s_{N_i} k) \\ \cos(\tilde{s}_i k) \end{bmatrix} \quad \text{and} \quad \boldsymbol{s}(k,i) = \begin{bmatrix} \sin(s_0 k) \\ \sin(s_1 k) \\ \vdots \\ \sin(s_{N_i} k) \\ \sin(\tilde{s}_i k) \end{bmatrix},$$

where $\tilde{s}_i$ is the position of the ith BPM. It is possible to calculate $G_b(k,i)$ for all values of i efficiently in one step. We consider therefor the construction of the vectors $\boldsymbol{r}[i]$, $\boldsymbol{c}(k,i)$

and $s(k,i)$, defining N_b and N_q as the total number of BPMs and quadrupoles and using $\tilde{s}_i$ for the BPM position. Then $G_b(k,i)$ is given by the ith element of the vector $\boldsymbol{G}_b(k)$, which is defined by

$$\boldsymbol{G}_b(k)^2 = \left[\tilde{\boldsymbol{R}}\tilde{\boldsymbol{c}}(k)\right]^{\circ 2} + \left[\tilde{\boldsymbol{R}}\tilde{\boldsymbol{s}}(k)\right]^{\circ 2} \quad \text{with} \tag{2.70}$$

$$\tilde{\boldsymbol{R}} = \begin{bmatrix} \tilde{\boldsymbol{r}} & \boldsymbol{R} & -\boldsymbol{I} \end{bmatrix}, \qquad \tilde{\boldsymbol{r}} = \begin{bmatrix} \tilde{r}_1 \\ \vdots \\ \tilde{r}_{N_b} \end{bmatrix},$$

$$\tilde{\boldsymbol{c}}(k) = \begin{bmatrix} \cos(s_0 k) \\ \cos(s_1 k) \\ \vdots \\ \cos(s_{N_q} k) \\ \cos(\tilde{s}_1 k) \\ \vdots \\ \cos(\tilde{s}_{N_b} k) \end{bmatrix} \quad \text{and} \quad \tilde{\boldsymbol{s}}(k) = \begin{bmatrix} \sin(s_0 k) \\ \sin(s_1 k) \\ \vdots \\ \sin(s_{N_q} k) \\ \sin(\tilde{s}_1 k) \\ \vdots \\ \sin(\tilde{s}_{N_b} k) \end{bmatrix},$$

where $^{\circ 2}$ symbolises the element-wise square of a vector also called Hadamard's square, $\boldsymbol{R}$ is the orbit response matrix and $\boldsymbol{I}$ is the identity matrix. Note that too large simulation noise (due to the limited number of used particles in the beam) in $\boldsymbol{R}$ can lead to numerical problems at the evaluation of expression Eq. (2.70).

2.2.3. Sensitivity function for Luminosity loss

The creation of a sensitivity function $G_{\Delta\mathcal{L}}(k)$ for the luminosity loss $\Delta\mathcal{L}$ due to ground motion is described in this section. This sensitivity function will be used to create a model, which will employed in this thesis for the design of the L-FB. Some use cases are shown in Sec. 3.3.

The sensitivity function $G_{\Delta\mathcal{L}}(k)$ is determined by numerical calculation (procedure already explained in Sec. 2.2.1.2) with the help of the simulation environment (Sec. 2.4). Modification to the given formulas have to be made however, since $\Delta\mathcal{L}$ is not linearly, but quadratically dependent on the ground motion. These quadratic dependence is valid for a luminosity loss smaller than 20 %, as has been verified via simulations.

The procedure to calculate $G_{\Delta\mathcal{L}}(k)$ has the following form. The beam line is misaligned sequentially with sine and cosine waves with growing, logarithmically distributed amplitudes $A(k)$. The first amplitude $A(k)$, which results in a luminosity loss $\mathcal{L}_{10\%}$ of more than 10 % is used for the calculation of $G_{\Delta\mathcal{L}}(k)$. The threshold of 10 % is chosen to ensure that the luminosity loss is large compared to the simulation noise of the luminosity calculation (about 1 %) and at the same time stays in the quadratic regime $\Delta\mathcal{L} \propto A(k)^2$. The values for $G_{\Delta\mathcal{L}}(k)$ can then be calculated by

$$G_{\Delta\mathcal{L}}(k) = \sqrt{\frac{\mathcal{L}_0 - \mathcal{L}_{\sin,10\%}(k)}{A_{\sin}(k)} + \frac{\mathcal{L}_0 - \mathcal{L}_{\cos,10\%}(k)}{A_{\cos}(k)}}, \tag{2.71}$$

where the indices sin und cos indicate the excitation with sine and cosine waves. Note that Eq. (2.71) is slightly different than the expression for a linear dependent observation

2. Modelling and simulation of ground motion effects

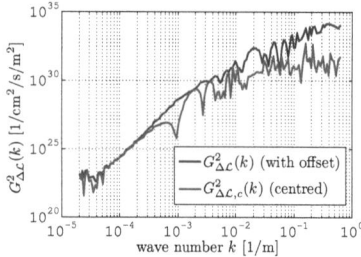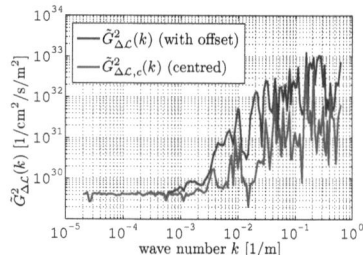

Figure 2.4.: Squared sensitivity functions $G^2_{\Delta\mathcal{L}}(k)$, $G^2_{\Delta\mathcal{L},c}(k)$ (left) and $\tilde{G}^2_{\Delta\mathcal{L}}(k)$, $\tilde{G}^2_{\Delta\mathcal{L},c}(k)$ (right), where $k = 2\pi/\lambda$ and λ is the wave length. The index c symbolises that the beams have been centred and the tilde-index that the final doublet (FD) has been stabilised. The non-stabilised sensitivity functions (left) are more sensitive to short wave lengths than the stabilised ones (right) due to the misalignment of the FD quadrupoles that result in beam-beam offset. On the other hand, the stabilised sensitivity functions do not decrease for long wave lengths as the non-stabilised ones due to dispersive effects created by the beam offset in the FD quadrupoles.

parameter in Eq. (2.59). Using $G_{\Delta\mathcal{L}}(k)$, the luminosity loss due to ground motion can be calculated as

$$\Delta\mathcal{L} = \frac{1}{(2\pi)^2} \int_{-\infty}^{+\infty} \int_{-\infty}^{+\infty} P(\omega, k) \left| S_{ST}(e^{j\omega T_{d,ST}}) \right|^2 \left| \hat{S}(e^{j\omega T_d}) \right|^2 G^2_{\Delta\mathcal{L}}(k) \mathrm{d}\omega \mathrm{d}k. \quad (2.72)$$

The calculated luminosity loss originates from two effects: beam-beam offset and beam size growth at the IP. The sensitivity function $G_{\Delta\mathcal{L}}(k)$ includes both effects. For some studies it is interesting to separate the two effects. The sensitivity function $G_{\Delta\mathcal{L}}(k)$ is split therefor into two components $G^2_{\Delta\mathcal{L}}(k) = G^2_{\Delta\mathcal{L},o}(k) + G^2_{\Delta\mathcal{L},c}(k)$, where $G^2_{\Delta\mathcal{L},o}(k)$ corresponds to the luminosity loss only due to beam-beam offset and $G^2_{\Delta\mathcal{L},c}(k)$ only due to beam size growth. The term $G_{\Delta\mathcal{L},c}(k)$ can be calculated according to Eq. (2.71) only that instead of $\mathcal{L}_{10\%}$ and $A(k)$ the values $\mathcal{L}_{c,10\%}$ and $A_c(k)$ have to be used. The $\mathcal{L}_{c,10\%}$ corresponds to the luminosity created by the same procedure as described above only that the two beams have been artificially centred. Since the two beam profiles $B_1(x,y)$ and $B_2(x,y)$ are not necessarily Gaussian (asymmetry, long tails), the subtraction of the beam centre positions is not a reliable way to align the beam centres, which are responsible for most of the luminosity. A better way two calculate proper beam offsets Δx and Δy is to maximise the expression

$$J(\Delta x, \Delta y) = \int_{-x_{max}}^{x_{max}} \int_{-y_{max}}^{y_{max}} B_1(x,y) B_2(x + \Delta x, y + \Delta y) \mathrm{d}x \mathrm{d}y, \quad (2.73)$$

with respect to Δx and Δy, where $J(\Delta x, \Delta y)$ is approximately proportional to the luminosity. Finally the $G_{\Delta\mathcal{L},o}(k)$ can be calculated by $G^2_{\Delta\mathcal{L},o}(k) = G^2_{\Delta\mathcal{L}}(k) - G^2_{\Delta\mathcal{L},c}(k)$. The resulting sensitivity functions are plotted in Fig. 2.4 (left).

When calculating $G_{\Delta\mathcal{L}}(k)$ the quadrupoles are misaligned exactly as the ground. In reality however, the ground waves first pass through the stabilisation systems on which the quadrupoles are located. Since two different stabilisation systems—quadrupole stabilisation and pre-isolator—are used, the motion of the quadrupoles is not a simple harmonic function anymore. Especially for high frequencies the pre-isolator is more efficient (see Fig. 1.8 (right)). Thus the last two quadrupoles (FD quadrupoles) are hardly influenced by ground motion compared to the other quadrupoles. This situation can be modelled by displacing all quadrupoles up to the FD quadrupoles with the usual sine and cosine waves, while the FD quadrupoles are not misaligned at all. The simulations described above can now be repeated for this new "stabilised" misalignment, which results in the sensitivity functions $\tilde{G}_{\Delta\mathcal{L}}(k)$ and $\tilde{G}_{\Delta\mathcal{L},c}(k)$ depicted in Fig. 2.4 (right).

Since the FD quadrupoles are very sensitive to beam-beam offset, $\tilde{G}_{\Delta\mathcal{L},o}(k)$ is smaller than $G_{\Delta\mathcal{L},o}(k)$, since the FD quadrupoles do not kick the beam. Additionally, $\tilde{G}_{\Delta\mathcal{L},c}(k)$ is now larger than $G_{\Delta\mathcal{L},o}(k)$. This is due to the fact that the beam comes onto the FD quadrupoles with an offset, which creates dispersion and other secondary effects. To investigate these effects quantitatively, a model will be presented in Sec. 2.3.2. Note that both, the stabilised and the non-stabilised sensitivity functions are idealisations of the reality. For low frequencies the non-stabilised and for high frequencies the stabilised sensitivity function is more representative. Simulations showed, however, that the use of the stabilised sensitivity function describes the reality more accurately.

2.2.4. Use of the models for L-FB design and performance prediction

With the help of the sensitivity function developed in Sec. 2.2.3 a luminosity loss model of the form Eq. (2.72) can be created. It can be extended to include also the sensor noise and the influence of the stabilisation system as

$$\Delta\mathcal{L} = \frac{1}{(2\pi)^2} \int_{-\infty}^{+\infty} \int_{-\infty}^{+\infty} \tilde{G}_{\Delta\mathcal{L}}^2(k) P(\omega, k) |S_{ST}(e^{j\omega T_{d,ST}})|^2 |S(e^{j\omega T_d})|^2 \\ + N(\omega) |C(e^{j\omega T_d}) S(e^{j\omega T_d})|^2 d\omega dk, \quad (2.74)$$

where $S_{ST}(e^{j\omega T_d})$ is the quadrupole motion frequency response of the stabilisation system to ground motion and $-C(e^{j\omega T_d})S(e^{j\omega T_d})$ is the quadrupole frequency response of the L-FB to BPM noise. The PSD $N(\omega)$ of the BPM noise is a flat spectrum (white noise) with a value according to Eq. (3.31) that is adjusted to the variance of the BPMs. The used frequency response of the L-FB is only dependent on ω, but not on the spatial form of the ground motion represented by the wave number k. This is a simplification to the real L-FB, which behaves differently for different spatial forms of the excitation. Even though, the model Eq. (2.74) delivers surprisingly good estimates. This is mainly due to the fact that the important ground motion components are all similarly treated by the L-FB. The model is used in this thesis for the design of the time-dependent part of the L-FB named $g(z)$ in Sec. 3.2.3.2 and for studies of cost reduction options in Sec. 3.3.

Even though model Eq. (2.74) is good for first estimates, it is not capable of covering the dependence of the behaviour of the system controlled by the L-FB, on the spatial form of the ground motion excitation (directionality) and can therefore not be used for a quantitive controller optimisation. In the following, a method that can be used for controller optimisation will be developed in two steps. Firstly, a straight forward approach will be presented. It will turn out however that this basic approach is not

2. Modelling and simulation of ground motion effects

capable to model the system behaviour properly. The analysis of the weak points of the basic method will lead to an improved approach that is based on the use of so called *virtually independent excitations*.

2.2.4.1. Basic approach

When trying to generalise the model a basic problem arises. While time-dependent harmonic functions are eigenfunctions of the accelerator system, spatial harmonic functions are no eigendirections of the system. This means that a sine excitation in time will be counter-acted by the L-FB with actuator settings, which are also a sine wave. The same is not true for spatial harmonic excitation. The spatial eigendirections of the controlled system are the columns of the input matrix V. This matrix is given by the singular value decomposition of the orbit response matrix $R = U\Sigma V^T$ (see Sec. 3.2.3.1 and App. C.4 for more explanation). It is therefore useful to describe the action of the L-FB with a frequency response $\hat{C}(e^{j\omega T_d}, i)$ given not only in its time-dependent eigenfunctions, but also its spatial eigendirections. With such a representation each eigendirection can be analysed independently, since it is decoupled from the other directions. The derivation of the frequency response matrices $\hat{C}(e^{j\omega T_d}, i)$ of the controller is straight forward, since the L-FB uses a decoupling into the eigendirections anyway (see Sec. 3.2.3.1 for a more detailed explanation). We use the hat index to symbolise decoupled systems and signals, where i is the index over the N_v eigendirections $v[i]$.

For a controller design, also the spectra of the excitation signals must be given in terms of $v[i]$ to allow a model based approach. The straight forward idea is now to project the ground motion waves of the spectrum $P(\omega, k)$ onto the matrix V to form a new spectrum $\hat{P}(\omega, i)$. As a next step, new luminosity sensitivity functions $G_{\Delta\mathcal{L}}[i]$ and $\tilde{G}_{\Delta\mathcal{L}}[i]$ can be calculated similar to the ones described in Sec. 2.2.3, with the difference that as excitations the input vectors $v[i]$ instead of the sine and cosine waves. A problem arises, since the $v[i]$ are only defined over one part of the linac (electron or positron part). It is not obvious how an excitation to both parts can be achieved. The solution is given in Sec. 3.2.3.3, where the according sensitivity functions $\tilde{G}_{\Delta\mathcal{L}}[i]$ (see Fig. 3.20) are used for a different purpose. With the created $\tilde{G}_{\Delta\mathcal{L}}[i]$, a new model of the form

$$\Delta\mathcal{L} = \frac{1}{(2\pi)^2} \int_{-\infty}^{+\infty} \sum_{i=1}^{N_v} \tilde{G}_{\Delta\mathcal{L}}[i]^2 \hat{P}(\omega, i) |S_{ST}(e^{j\omega T_d})|^2 |\hat{S}(e^{j\omega T_d}, i)|^2 d\omega \qquad (2.75)$$

can be created, where also the BPM noise is neglected here for sake of simplicity.

When this modelling approach is used for the luminosity predictions, the results are unfortunately wrong. The reason why this transformation of the problem is not able to represent reality accurately is visualised in Fig. 2.5. A harmonic wave with long wave length is known to result in a small $\Delta\mathcal{L}$ due to its smooth nature. When this wave is split up into the vectors $v[i]$ as $\sin(s[i]k + \phi) = \sum_i a[i]v[i]$, the individual components $a[i]v[i]$ can be much more jagged than the original wave. The application of the $v[i]$ results in a much larger $\Delta\mathcal{L}$, if they are assumed to be independent of each other. Mathematically this simply means that the luminosity loss is not a linear function of the excitation, since

2.2. Luminosity and beam orbit impact of ground motion

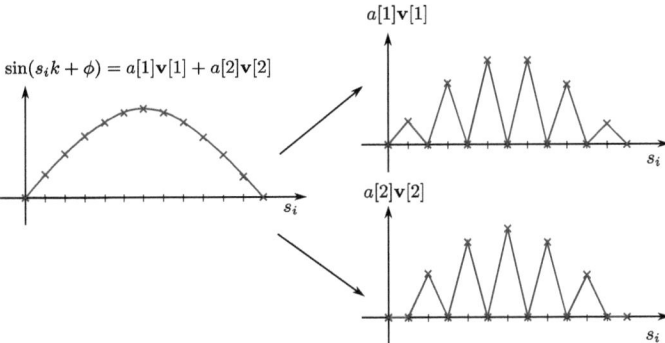

Figure 2.5.: A ground motion wave $\sin(s_i k + \phi)$ is projected onto two eigendirections $\boldsymbol{v}[1]$ and $\boldsymbol{v}[2]$, which fully describe the original wave. The two eigendirections are much more jagged than the original wave and create therefore a much bigger luminosity loss, if they are applied independently. The PSD of the projected motion can thus not be used for a luminosity prediction, since its individual components are assumed to be uncorrelated.

the additivity is not given.

$$\Delta \mathcal{L} \left\{ \sin(s[i]k + \phi) \right\} = \Delta \mathcal{L} \left\{ \sum_i a[i] \boldsymbol{v}[i] \right\}$$
$$\neq \Delta \mathcal{L} \left\{ a[1] \boldsymbol{v}[1] \right\} + \cdots + \Delta \mathcal{L} \left\{ a[N_v] \boldsymbol{v}[N_v] \right\} \qquad (2.76)$$

In reality the elements of $\hat{P}(\omega, i)$ are not independent of each other, but are created from the elements of $P(\omega, k)$. In Eq. (2.75) independence of the elements of $\hat{P}(\omega, i)$ is assumed however, which causes the vast overestimation.

2.2.4.2. Approach using virtually independent excitations

To resolve the problem discovered in the last section, we use the following modified approach. The ground motion components of the different decoupled controller channels are not independent. We however need such an independent description for the developed modelling framework. Therefore, we calculate a *virtually independent excitation* $\hat{P}_v(\omega, i)$, which creates exactly the same beam oscillation spectrum $\hat{B}(\omega, i)$ as the real ground motion spectrum $P(\omega, k)$. The index i in $\hat{B}(\omega, i)$ again corresponds to the ith decoupled loop, which is associated with the output eigendirections $\boldsymbol{u}[i]$ (columns of $\boldsymbol{U}$). The decoupling procedure is explained in more detail in Sec. 3.2.3.1. The physical interpretation of the $\boldsymbol{u}[i]$ is that an excitation of the quadrupoles with $\boldsymbol{v}[i]$ results in a beam oscillation $s(i)\boldsymbol{b}[i]$, where $s[i]$ is the according singular value.

Two remarks should be made to this approach. First, the technique of using a virtual excitation spectrum focuses on a correct representation of the beam oscillations, while the quadrupole motion is not modelled correctly. This approach represents reality very well, since the L-FB is much more a beam oscillation damping feedback than a quadrupole

2. Modelling and simulation of ground motion effects

alignment feedback. The corrections applied by the L-FB actuators to the quadrupole motion are very small compared to the ground motion displacement. The L-FB adds just a small signal to damp the beam oscillations. The picture that the L-FB aligns the quadrupole is hence wrong and a focusing on the absolute value of the quadrupole displacement unimportant.

Secondly, it should be mentioned that the elements of $\hat{B}(\omega, i)$ are all independent, since the elements of $\hat{P}(\omega, i)$ are independent. This is not the case in reality however, where the elements of the beam oscillation spectrum are correlated. We neglect this fact for the controller design for sake of simplicity.

The PSD $\hat{P}_v(\omega, i)$ can be easily found from a given $\hat{B}(\omega, i)$ by

$$\hat{P}_v(\omega, i) = s[i]^{-1} \hat{B}(\omega, i), \qquad (2.77)$$

since the excitation $v[i]$ and the beam oscillation $b[i]$ of one decoupled channel are simply connected by the according singular value $s[i]$. The creation of $\hat{B}(\omega, i)$ is very similar to the one for the not projected oscillations $B(\omega, i)$, for which a sensitivity function was derived in Sec. 2.2.2. Only Eq. (2.70) has to be modified slightly to

$$\hat{G}_b(k)^2 = \left[\boldsymbol{U}^T \tilde{\boldsymbol{R}} \tilde{\boldsymbol{c}}(k)\right]^{\circ 2} + \left[\boldsymbol{U}^T \tilde{\boldsymbol{R}} \tilde{\boldsymbol{s}}(k)\right]^{\circ 2}. \qquad (2.78)$$

To clarify the notation, the index i of $\hat{G}_b(k, i)$ corresponds to the ith element of the vector on the right side. The projected spectrum is finally calculated by

$$\hat{B}(\omega, i) = \frac{1}{2\pi} \int_{-\infty}^{+\infty} P(\omega, k) \left|S_{ST}(e^{j\omega T_{d,ST}})\right|^2 \hat{G}_b(k, i)^2 \mathrm{d}k, \qquad (2.79)$$

where $S_{ST}(e^{j\omega T_{d,ST}})$ is the frequency response of the quadrupole stabilisation. The spectrum $\hat{P}_v(\omega, i)$ is the key element for the L-FB optimisation in Sec. 3.2.3.3. Even though $\hat{P}_v(\omega, i)$ has proven to be very useful for the prediction of beam oscillations, it cannot be used for luminosity loss prediction, since the correlation of the different signals is not accurately modelled. For luminosity loss prediction, still the simple model Eq. (2.74) has to be used.

2.3. Effects due to the final doublet offsets

2.3.1. Beam-beam offset

The beam-beam offset at the IP is very sensitive to the misalignment of the last two quadrupoles of the beam line (QD0 and QF1) forming the so called final doublet (FD). To isolate these quadrupoles from the ground a dedicated mitigation method was designed. The so called pre-isolator is a massive concrete block supported by ten pneumatic vibration isolators. A picture of the system is plotted in Fig. 2.6. While the structure of the pre-isolator was already covered in Sec. 1.3.2.5, we focus here on the derivation of a model describing its influence on the beam-beam offset.

When deriving such a model, it is not enough to only consider vertical and horizontal displacements. Since the concrete block has an longitudinal extension of about 8 m, also the tilte mode (in beam travel direction) has to be included in the model. For this

2.3. Effects due to the final doublet offsets

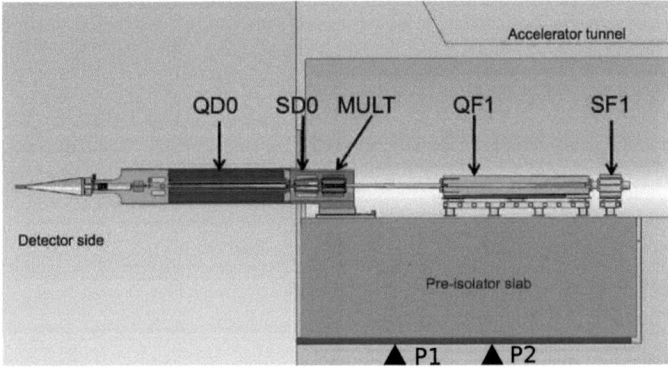

Figure 2.6.: Side view of the final doublet system, which is located just in front of the detector hall (IP to the left side of the plot). The system consists of (from left to right) the quadrupole magnet QD0, the sextupole magnet SD0, a octupole magnet MULT, the quadrupole magnet QF1 and the sextupole magnet SF1. A lever arm construction is used to position QD0 as close as possible to the IP. The magnets are placed on the pre-isolator in order to suppress ground motion. The support of the pre-isolator is simplified in this picture and modeled by only two contact points P1 and P2. Picture courtesy of F. Ramos, where the support points have been added to the original plot.

reason, the pre-isolator design team performed measurements to determine the frequency response functions of the displacements of QD0 and QF1 with respect to an excitation of respectively the four front and the four back vibration isolators. Even though these two excitations act in a distributed manner on the pre-isolator, they are modelled to be point-like (in points $P1$ and $P2$). The distances of these two points to the IP are named s_1 and s_2. The mentioned frequency response functions are published in Gaddi [43] and are referred to as $H_{s_1 \to 0}(j\omega)$, $H_{s_1 \to 1}(j\omega)$, $H_{s_2 \to 0}(j\omega)$ and $H_{s_2 \to 1}(j\omega)$, where e.g. the term $H_{s_1 \to 0}(j\omega)$ corresponds to the frequency response function from point $P1$ to QD0.

Using these informations a sensitivity function for the beam-beam offset $G_\delta(\omega, k)$ can be derived as follows. A travelling wave $\sin(sk - \omega t + \phi)$ can be split up into two standing waves of the same amplitude; one sine and one cosine like with respect to the IP (see Eq. (2.57)). A displacement of the FD with a cosine wave does not create any beam-beam offset, since both beams are kicked by QD0 and QF1 exactly the same way. We can restrict ourselves hence, to analyse the effect of a sine wave. Such an excitation displaces the support points with $\sin(s_1 k)$ and $\sin(s_2 k)$. The transmission of these motions to the quadrupoles can be described by frequency response functions. The motion of e.g. QD0 called y_0 can thus be written in the frequency space as

$$y_0(j\omega, k) = H_{s_1 \to 0}(j\omega)\sin(s_1 k) + H_{s_2 \to 0}(j\omega)\sin(s_2 k). \qquad (2.80)$$

The quadrupole motion $y_0(j\omega, k)$ results in a beam displacement at the IP of $y_{IP} = r_0 y_0$, where r_0 is defined as $r_0 = \mathrm{d}y_{IP}/\mathrm{d}y_0$. Applying Eq. (2.80) also to QF1 and considering that the beam-beam offset y_δ for a sine wave is twice the individual beam displacement

47

2. Modelling and simulation of ground motion effects

y_{IP} (since there are two beams), the sensitivity function for the beam-beam offset can be written as

$$G_\delta(j\omega, k) = 2r_0 \left[H_{s_1 \to 0}(j\omega) \sin(s_1 k) + H_{s_2 \to 0}(j\omega) \sin(s_2 k) \right]$$
$$+ 2r_1 \left[H_{s_1 \to 1}(j\omega) \sin(s_1 k) + H_{s_2 \to 1}(j\omega) \sin(s_2 k) \right] \quad \text{with} \quad (2.81)$$
$$r_0 = 4/3, \quad s_1 = 8.306 \, \text{m}$$
$$r_1 = -1/3, \quad \text{and} \quad s_2 = 10.946 \, \text{m}.$$

Note that this sensitivity function is also a function of ω, which is in contrast to the previously derived sensitivity functions. The variance of the beam-beam offset $\sigma_\delta^2 = \mathbb{E}\{y_\delta^2\}$ can finally be calculated by

$$\sigma_\delta^2 = \frac{1}{(2\pi)^2} \int_{-\infty}^{\infty} \int_{-\infty}^{\infty} P(\omega, k) \left| G_\delta(\omega, k) \right|^2 \left| S_{IP}(e^{j\omega T_d}) \right|^2 d\omega dk, \quad (2.82)$$

where $S_{IP}(e^{j\omega T_d})$ is the frequency response function of the IP-FB to ground motion. The model Eq. (2.82) will be used in Sec. 3.4 for the optimisation of the IP-FB.

2.3.2. Beam size growth

The offsets of the final doublet magnets QD0 and QF1 are not only causing beam-beam offset at the IP. Full-scale simulations with the simulation framework in Sec. 2.4 indicated also correlation between the magnet misalignments and the beam size growth at the IP. To understand this effect, a model will be developed in this section, which allows to predict the associated luminosity loss.

The estimation of the beam size growth at the IP is based on a two-particle model of the CLIC beam. Two macro particles are used to represent the positive and negative standard deviation of the energy of the Gaussian CLIC beam. After tracking these particles by the transfer matrix formalism to the IP, their offsets are used to estimate the beam size growth. We will in the following focus on the vertical direction, since the beam size growth in the horizontal direction is negligible.

The motion of one particle, with a position y, angle y' and energy E, through the final focus system is mainly determined by the two quadrupole magnets QD0 and QF1, the two sextupole magnets SD0 and SF1 and the drifts (empty space) between these elements (see Fig. 2.6). The multipole in front of SD0 can be neglected for our purposes. The following expressions describe the particle transport through these elements and can be taken from Wille [136] and Wiedemann [135].

The motion of a particle, with coordinates $\boldsymbol{y}_i = [y, y']^T$ through a drift element is given by

$$\boldsymbol{y}_o = \begin{bmatrix} 1 & L_D \\ 0 & 1 \end{bmatrix} \boldsymbol{y}_i = \boldsymbol{M}_D \boldsymbol{y}_i, \quad (2.83)$$

where $\boldsymbol{y}_o$ corresponds to the particle coordinates after the drift space and L_D is the length of the drift space. For focusing and defocusing quadrupoles QF and QD the

2.3. Effects due to the final doublet offsets

particle transfer can be described by

$$\boldsymbol{y}_{o,F} = \begin{bmatrix} \cos\left(\sqrt{k_{QF}}L_{QF}\right) & \frac{1}{\sqrt{k_{QF}}}\sin\left(\sqrt{k_{QF}}L_{QF}\right) \\ -\sqrt{k_{QF}}\sin\left(\sqrt{k_{QF}}L_{QF}\right) & \cos\left(\sqrt{k_{QF}}L_{QF}\right) \end{bmatrix} \boldsymbol{y}_i = \boldsymbol{M}_{QF}\boldsymbol{y}_i \quad (2.84)$$

$$\boldsymbol{y}_{o,D} = \begin{bmatrix} \cosh\left(\sqrt{k_{QD}}L_{QD}\right) & \frac{1}{\sqrt{k_{QD}}}\sinh\left(\sqrt{k_{QD}}L_{QD}\right) \\ \sqrt{k_{QD}}\sinh\left(\sqrt{k_{QD}}L_{QD}\right) & \cosh\left(\sqrt{k_{QD}}L_{QD}\right) \end{bmatrix} \boldsymbol{y}_i = \boldsymbol{M}_{QD}\boldsymbol{y}_i, \quad (2.85)$$

where k_Q is the quadrupole strength. To track particles with different energies, it is useful to express the quadrupole strength as $k_Q = \tilde{k}_Q/E$, where $\tilde{k}_Q$ does not depend on the particle engergy E. The use of the thin lens approximation for QD0 and DF1 is not possible for the final doublet magnets, since it would cause too large approximation errors. Note that QD0 is a focusing and QF1 a defocusing magnet for the vertical plane, since the naming of these magnets is according to their function in the horizontal plane. The effect of an offset y_Q of a quadrupole can be modelled by moving instead of the quadrupole, the beam into the opposite direction, transferring the beam through the quadrupole and moving the beam back into its initial coordinate frame (see Schulte [112]). If the vector $\boldsymbol{y}_Q = [y_Q\ 0]^T$ is defined, this procedure is expressed mathematically by

$$\boldsymbol{y}_o = \boldsymbol{M}_Q\left(\boldsymbol{y}_i - \boldsymbol{y}_Q\right) + \boldsymbol{y}_Q = \boldsymbol{M}_Q\boldsymbol{y}_i + (\boldsymbol{I} - \boldsymbol{M}_Q)\boldsymbol{y}_Q. \quad (2.86)$$

Differently than drifts and quadrupoles, sextupoles are non-linear elements. The vertical kick from a sextupole over a short distance is

$$\Delta y' = \frac{2\tilde{k}_S}{E}xy, \quad (2.87)$$

where $\tilde{k}_S$ is the integrated sextupole strength and x the beam position in the horizontal direction. We model the sextupoles SD0 and SF1 by tracking the beam to the centre of the sextupole by a drift space, applying the kick in Eq. (2.87) and tracking the modified particle over a second drift space, also with the half length of the sextupole. When applying the kick of Eq. (2.87), the problem arises that the value of x is unknown. We thus use instead of the unknown x its expectation value. For this reason x is split up in the betatron and the dispersive motion $x = x_\beta + x_D$. This results in

$$\mathbb{E}\left\{x\right\} = \mathbb{E}\left\{x_\beta\right\} + \mathbb{E}\left\{x_D\right\} = d_x\delta_E, \quad (2.88)$$

where d_x is the dispersion in the horizontal direction as defined in Eq. (A.9) and δ_E is the relative energy deviation of the particle $\delta_E = (E - E_0)/E_0$, where E_0 is the mean energy of all beam particles. The transport matrix of a sextupole is hence given by

$$\boldsymbol{y}_o = \begin{bmatrix} 1 & L_S/2 \\ 0 & 1 \end{bmatrix} \begin{bmatrix} 1 & 0 \\ \frac{2\tilde{k}_s}{E}d_x\delta_E & 1 \end{bmatrix} \begin{bmatrix} 1 & L_S/2 \\ 0 & 1 \end{bmatrix} \boldsymbol{y}_i. \quad (2.89)$$

Using the expressions Eqs. (2.83), (2.84), (2.85), (2.86) and (2.89), the beam transport of one particle through the final doublet is given by

$$\begin{aligned} \boldsymbol{y}_1 &= \boldsymbol{M}_{QF1}\boldsymbol{M}_{D1}\boldsymbol{M}_{SF1}\boldsymbol{y}_i + (\boldsymbol{I} - \boldsymbol{M}_{QF1})\boldsymbol{y}_{QF1} \\ \boldsymbol{y}_2 &= \boldsymbol{M}_{QD0}\boldsymbol{M}_{D3}\boldsymbol{M}_{SD0}\boldsymbol{D}_2\boldsymbol{y}_1 + (\boldsymbol{I} - \boldsymbol{M}_{QD0})\boldsymbol{y}_{QD0} \\ \boldsymbol{y}_{IP} &= \boldsymbol{M}_{D4}\boldsymbol{y}_2, \end{aligned} \quad (2.90)$$

2. Modelling and simulation of ground motion effects

where y_i is the particle before SF1, y_1 after QF1, y_2 after QD0 and y_{IP} at the IP. Misalignments of sextupoles are not considered in this model. Since only the beam size growth at the IP is of interest, only the first component of y_{IP} has to be evaluated. This gives an expression of the form

$$y_{IP} = T_1(E)y_i + T_2(E)y_i' + T_3(E)y_{QF1} + T_4(E)y_{QD0}. \qquad (2.91)$$

The coefficients $T_1(E)$ to $T_4(E)$ of this linear relation are only dependent on the particle energy E and can be obtained by evaluating Eq. (2.90).

Using two macro particles, both with $y_i = 0$ and $y_i' = 0$, but with different energies $E_0 - \sigma_E$ and $E_0 + \sigma_E$, where σ_E the standard deviation of the particle energies, the beam size growth due to dispersion at the IP can be modelled using Eq. (2.91) by

$$\Delta\sigma_{y,disp}^* = \frac{y_{IP}(E_0 - \sigma_E) - y_{IP}(E_0 + \sigma_E)}{2}$$
$$= \frac{T_3(E_0 - \sigma_E) - T_3(E_0 + \sigma_E)}{2}y_{QF1} + \frac{T_4(E_0 - \sigma_E) - T_4(E_0 + \sigma_E)}{2}y_{QD0}. \qquad (2.92)$$

Considering that

$$\Delta\sigma_y^* = \sqrt{\left(\sigma_{y,core}^*\right)^2 + \left(\Delta\sigma_{y,disp}^*\right)^2} - \sigma_{y,core}^*,$$

and using the approximations in Eqs. (A.3) and (A.8), the beam size growth can be converted to an associated relative luminosity loss as

$$\frac{\Delta \mathcal{L}}{\mathcal{L}_0} \approx \frac{\Delta\sigma_y^*}{\sigma_{y,core}^*} \approx \left(\frac{\Delta\sigma_{y,disp}^*}{\sigma_{y,core}^*}\right)^2, \qquad (2.93)$$

where $\mathcal{L}_0$ is the nominal luminosity and we use for the core beam size $\sigma_{y,core}^* = 0.77$ nm. We define the core beam size as the distance from the beam centre, where the beam histogram is reduced to 60.6% of its maximum value. If the beam would be purely Gaussian this value would be equal to the standard deviation. Since the beam is not purely Gaussian, the standard deviation of the beam distribution $\sigma_{y,0}^* = 1.41$ nm is larger than $\sigma_{y,core}^*$. For the luminosity prediction the core beam size is more relevant and is thus used in Eq. (2.93). The created luminosity model is compared to simulations results in Fig. 2.7.

While the dispersive beam size growth explains the luminosity loss due to an offset of QD0 very well, a second effect has to be considered when modelling the offset of QF1. When estimating the coupling from the horizontal to the vertical direction in Eq. (2.88) only first order effects have been considered so far. An offset of the beam in SD0 y_{SD0} due to a kick in QF1 results however also in a second order effect that can be estimated as

$$\mathbb{E}\left\{(\Delta y_{SD0}')^2\right\} = \left(\frac{2K_{SD0}}{E}\right)^2 y_{SD0}^2 \mathbb{E}\left\{x_{\beta,SD0}^2 + 2x_{\beta,SD0}x_{D,SD0} + x_{D,SD0}^2\right\}$$
$$= \left(\frac{2K_{SD0}}{E}\right)^2 y_{SD0}^2 \sigma_{x,SD0}^2 + \left(\frac{2K_{SD0}}{E}\right)^2 y_{SD0}^2 (d_{x,SD0}\Delta E)^2, \qquad (2.94)$$

2.3. Effects due to the final doublet offsets

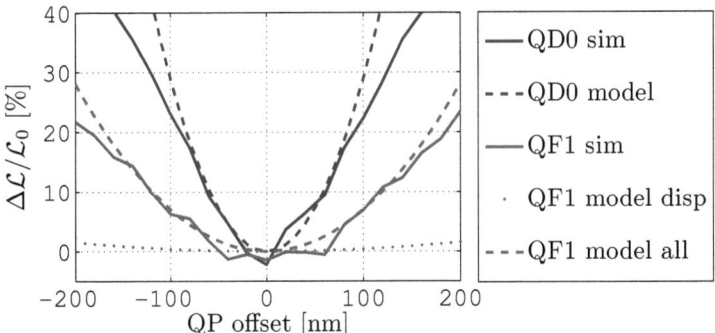

Figure 2.7.: Simulation results and models for the relative luminosity loss $\Delta\mathcal{L}/\mathcal{L}_0$ due to misalignments of QF1 and QD0. For the simulations synchrotron radiation was not considered, but it does not change to overall picture significantly. The simulated luminosity loss due to an offset of QD0 (blue curve) can be very well explained by the dispersive beam size growth model (dashed blue curve) given by Eq. (2.93). For an offset of QF1 (red curve), additionally to the dispersion (dotted red curve) also uncancelled geometric aberrations (dashed red curve) have to be considered according to Eq. (2.96). The final model describes the simulated luminosity loss very well, for relative losses smaller than 15 to 20 %.

where we assumed that the x_β and x_D are uncorrelated and σ_x is the horizontal beam size in the centre of the sextupole SD0 without dispersion. The second term in Eq. (2.94) corresponds to the dispersion, which is coupled from the horizontal into the vertical direction. This effect has been already considered in the dispersion model Eq. (2.92). The first term of Eq. (2.94) corresponds to uncorrected geometric aberrations. Such kicks result in a beam size growth at the IP, which can be estimated with the help of second order transport matrices (see Napoly [76] and Brown [16]) as

$$(\Delta\sigma^*_{y,aberr})^2 = m_{1,2}^2 \left(\frac{2K_{SD0}}{E}\right)^2 y_{SD0}^2 \sigma_{x,SD0}^2, \qquad (2.95)$$

where $m_{1,2}$ is the element of the first row and second column of the matrix

$$M_{aberr} = M_{D4} M_{QD0} M_{D3} \begin{bmatrix} 1 & L_{SD0}/2 \\ 0 & 1 \end{bmatrix}.$$

Considering also $\Delta\sigma^*_{IP,aberr}$ for the luminosity estimate, we get the final expression

$$\frac{\Delta\mathcal{L}}{\mathcal{L}_0} \approx \left(\frac{\Delta\sigma^*_{y,disp}}{\sigma^*_{y,core}}\right)^2 + \left(\frac{\Delta\sigma^*_{y,aberr}}{f_y \sigma^*_{y,core}}\right)^2 \quad \text{with} \quad f_y = \frac{\sigma^*_{y,0}}{\sigma^*_{y,core}}. \qquad (2.96)$$

The scaling factor f_y accounts for the fact that the beam size growth due to geometric aberrations corresponds to the overall, non Gaussian beam, which has a standard

2. Modelling and simulation of ground motion effects

deviation of $\sigma_{y,0}^* = 1.41\,\text{nm}$. Only the relative beam size growth of the core beam is of importance for the luminosity loss, and hence the overall beam size growth due to geometric aberrations has to be scaled to the core beam size growth as $\Delta\sigma_{y,aberr}^*/f_y$. The model in Eq. (2.96) explains the luminosity loss due to the offsets of QF1 and QD0 very well for a relative loss up to 15 to 20 % as can be seen in Fig. 2.7.

2.4. Integrated simulation framework

2.4.1. Overview

The ultimate goal for the design of CLIC is (besides the energy) to reach the specified luminosity. The dependence of the luminosity on the ground motion is a complex non-linear system with a high number of input parameters. Even though the presented models in this chapter provide some estimates of individual effects due to ground motion, there is no single, easy to handle model that covers all influences accurately. Especially the complex interaction of the four ground motion mitigation methods is difficult to analyse.

Due to this reason, a framework was developed to perform detailed numerical simulations of the overall system (see Snuverink et al. [125]). This simulation framework combines and extends existing numerical simulation codes and tools and serves multiple purposes. First of all, the framework acts as a development and performance evaluation tool for the four mitigation methods. Even though these methods are designed with simpler analytic models (as presented in this chapter), the final verification and optimisation has to be performed with the computational more expensive simulation framework. It is used in this thesis in Chap. 4 for the performance evaluation of the designs of the L-FB and IP-FB. As a second task, the framework also serves as a final performance analysis tool, for not only the individual mitigation methods, but also for their interaction and the overall system performance by providing realistic luminosity values. These data should show to the accelerator community that CLIC can be operated despite of ground motion.

The integrated simulation framework combines four individual simulation tools. The beam tracking code PLACET (Schulte et al. [113]) tracks the particle beams through the main linac and BDS to the IP of CLIC. A ground motion generator (Renier, Bambade and Sery [102]) creates realistic element displacements for the beam tracking with PLACET. GUINEA-PIG (Schulte [108]) uses the tracked beams to calculate collision parameters as the luminosity. Finally the numerical tool Octave [34] (open source Matlab clone) serves as an interface between the other modules, handles the data in- and output and is used for the controller implementation. An overview of the whole framework is given in Fig. 2.8. In the following, the individual simulation codes are explained more in detail. In App. B the most important information about the installation, usage and input parameters of the framework are collected.

2.4.2. Individual components

PLACET: PLACET stands for "Program for Linear Accelerator Correction Efficiency Tests" and is a particle tracking code. It calculates the behaviour of particle beams

2.4. Integrated simulation framework

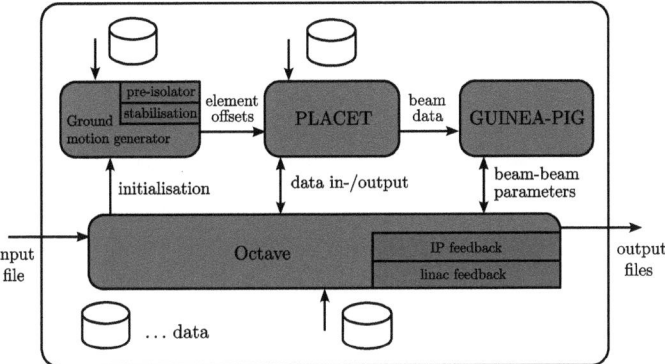

Figure 2.8.: The overall simulation framework consists of four modules: PLACET for the beam tracking, GUINEA-PIG for the simulation of beam-beam effects, the ground motion generator and Octave. Octave is the central steering element that controls the other three modules and handles the user in- and output. The ground motion mitigation methods (in red) are implemented either as an integral element of the ground motion generator or with the comfortable high-level functions of Octave.

along an accelerator. PLACET is especially well suited to take into account effects of accelerator component imperfections, as e.g. misalignment due to ground motion. In the integrated simulation framework, PLACET is used to track beams through the main linac and the BDS of the electron and positron part of CLIC. The code is written in C++ and is controlled usually with the script language Tcl/Tk. Additionally, it is possible to start an Octave environment within the execution of a PLACET simulation. This is a very comfortable feature, since the high-level functions of Octave can be used to implement mathematical algorithms and perform input and output operations.

Ground motion generator: In order to create realistic accelerator component displacements for the particle tracking in PLACET, a ground motion generator has been integrated into PLACET. It is a modified version of an already existing generator, which implements the strategy described in Sec. 2.1.2.4 for the creation of random fields with a given PSD. With the modifications, filter functions can now be assigned to accelerator components, which are used to filter the initial ground motion PSD. This feature allows to include the effects of the quadrupole stabilisation system and the pre-isolator in the simulations. The generator can produce displacements corresponding to the ATL law and the ground motion models A, B, B10 and C (see Sec. 2.1.2.4). The extension to other models is straight forward, if the according PSD is given. In the following, we will point out some deviations of the implemented ground motion generator from the standard procedure to create random fields introduced in Sec. 2.1.2.4.

In the current implementation not only one (as in Eq. (2.41)), but two travelling waves are used, which move in opposite direction to create a more realistic model. The ground movement is calculated at certain discrete times t_m and positions s_n (where m and n

2. Modelling and simulation of ground motion effects

are indices over all discrete positions and times) by

$$x_{ij}[t_m, s_n] = \frac{a_{ij}}{\sqrt{2}} \left[\sin\left[\omega_i t_m + k_j s_n + \phi_{ij}^{(1)}\right] + \sin\left[\omega_i t_m - k_j s_n + \phi_{ij}^{(2)}\right] \right] \quad (2.97)$$

Each of the two waves is carrying only half the power of the wave in Eq. (2.41). The superposition of the two waves forms one standing wave. To calculate the sum of all x_{ij} more efficiently Eq. (2.97) can be split up by using trigonometric identities into

$$\begin{aligned} x_{ij}[t_m, s_n] &= \frac{a_{ij}}{\sqrt{2}} \cos\left[k_j s_n\right] \left[\cos[\omega_i t_m] \left[\cos\phi_{ij}^{(1)} + \cos\phi_{ij}^{(2)}\right] - \sin[\omega_i t_m] \left[\sin\phi_{ij}^{(1)} + \sin\phi_{ij}^{(2)}\right] \right] \\ &\quad - \frac{a_{ij}}{\sqrt{2}} \sin\left[k_j s_n\right] \left[\cos[\omega_i t_m] \left[\sin\phi_{ij}^{(1)} - \sin\phi_{ij}^{(2)}\right] + \sin[\omega_i t_m] \left[\cos\phi_{ij}^{(1)} - \cos\phi_{ij}^{(2)}\right] \right] \\ &= \cos\left[k_j s_n\right] A_{ij}[t_m] - \sin\left[k_j s_n\right] B_{ij}[t_m] \end{aligned} \quad (2.98)$$

with

$$A_{ij}[t_m] = \frac{a_{ij}}{\sqrt{2}} \cos[\omega_i t_m] \left[\cos\phi_{ij}^{(1)} + \cos\phi_{ij}^{(2)}\right] - \frac{a_{ij}}{\sqrt{2}} \sin[\omega_i t_m] \left[\sin\phi_{ij}^{(1)} + \sin\phi_{ij}^{(2)}\right],$$

$$B_{ij}[t_m] = \frac{a_{ij}}{\sqrt{2}} \cos[\omega_i t_m] \left[\sin\phi_{ij}^{(1)} - \sin\phi_{ij}^{(2)}\right] + \frac{a_{ij}}{\sqrt{2}} \sin[\omega_i t_m] \left[\cos\phi_{ij}^{(1)} - \cos\phi_{ij}^{(2)}\right].$$

The signal $x[t_m, s_n]$ can now be efficiently computed at some discrete positions s_n and times t_m as

$$\begin{aligned} x[t_m, s_n] &= \sum_i \sum_j x_{ij}[t_m, s_n] = \sum_i \sum_j \cos[k_j s_n] A_{ij}[t_m] - \sin[k_j s_n] B_{ij}[t_m] \\ &= \sum_j \left[\cos[k_j s_n] \left[\sum_i A_{ij}[t_m]\right] \right] - \sum_j \left[\sin[k_j s_n] \left[\sum_i B_{ij}[t_m]\right] \right] \\ &= \sum_j \left[\cos[k_j s_n] \hat{A}_j[t_m] - \sin[k_j s_n] \hat{B}_j[t_m] \right] \quad \text{with} \end{aligned} \quad (2.99)$$

$$\hat{A}_j[t_m] = \sum_i A_{ij}[t_m] \quad \text{and} \quad \hat{B}_j[t_m] = \sum_i B_{ij}[t_m]. \quad (2.100)$$

Note that the terms $\hat{A}_j[t_m]$ and $\hat{B}_j[t_m]$ only have to be calculated once per time step. This reduces the computational complexity significantly.

The application of a ground motion filter of the form $F[j\omega_i] = A_F[\omega_i] e^{j\phi_F[\omega_i]}$ to Eq. (2.97), where j is here the imaginary unit and not the index over the wave numbers k, corresponds to a change of the harmonic functions to

$$\begin{aligned} x_{ij,F}[t_m, s_n] &= A_F[\omega_i] \frac{a_{ij}}{\sqrt{2}} \sin\left[\omega_i t_m + k_j s_n + \phi_{ij}^{(1)} + \phi_F[\omega_i]\right] \\ &\quad + A_F[\omega_i] \frac{a_{ij}}{\sqrt{2}} \sin\left[\omega_i t_m - k_j s_n + \phi_{ij}^{(2)} + \phi_F[\omega_i]\right]. \end{aligned} \quad (2.101)$$

Hence, the filtered ground motion can be calculated similarly as in Eqs. (2.99) and (2.100), with the difference that the expression for $A_{ij}[t_m]$ and $B_{ij}[t_m]$ have to be exchanged with

2.4. Integrated simulation framework

$A_{ij,F}[t_m]$ and $B_{ij,F}[t_m]$, which are given by

$$A_{ij,F}[t_m] = A_F[\omega_i]\frac{a_{ij}}{2}\cos[\omega_i t_m]\left[c_{ij}^{(1)} + c_{ij}^{(2)}\right] - A_F[\omega_i]\frac{a_{ij}}{\sqrt{2}}\sin[\omega_i t_m]\left[s_{ij}^{(1)} + s_{ij}^{(2)}\right],$$

$$B_{ij,F}[t_m] = A_F[\omega_i]\frac{a_{ij}}{\sqrt{2}}\cos[\omega_i t_m]\left[s_{ij}^{(1)} - s_{ij}^{(2)}\right] + A_F[\omega_i]\frac{a_{ij}}{\sqrt{2}}\sin[\omega_i t_m]\left[c_{ij}^{(1)} - c_{ij}^{(2)}\right] \quad \text{with}$$

$$c_{ij}^{(l)} = \cos\left[\phi_{ij}^{(l)} + \phi_F[\omega_i]\right] = \cos\left[\phi_{ij}^{(l)}\right]\cos[\phi_F[\omega_i]] - \sin\left[\phi_{ij}^{(l)}\right]\sin[\phi_F[\omega_i]], \quad (2.102)$$

$$s_{ij}^{(l)} = \sin\left[\phi_{ij}^{(l)} + \phi_F[\omega_i]\right] = \sin\left[\phi_{ij}^{(l)}\right]\cos[\phi_F[\omega_i]] + \cos\left[\phi_{ij}^{(l)}\right]\sin[\phi_F[\omega_i]]. \quad (2.103)$$

The splitting up of the sine and cosine of the sum of angles in Eqs. (2.102) and (2.103) is necessary, since only the sine and cosine values of the initial phases are stored due to implementation reasons.

GUINEA-PIG: After PLACET has produced realistic predictions for the electron and positron beams at the IP, they can be used by GUINEA-PIG to calculate the luminosity and other parameters of the according beam collisions. GUINEA-PIG stands for "Generator of Unwanted Interactions for Numerical Experiment Analysis—Program Interface to GEANT" and is a simulation code that models the complex physical processes involved in the beam collisions. To determine the mutual beam influence, the electro-magnetic fields of the highly relativistic electron and positron beams are calculated. Additionally, relevant effects of quantum electrodynamics and quantum chromodynamic (QCD) are taken into account. GUINEA-PIG can be used to calculate many different quantities connected to the beam-beam interaction as luminosity, deflection angle of the beams and the particle production for the most relevant background processes. Therefore, it can be used for the design of the interaction region and the detector itself.

Octave: Octave is an open-source Matlab clone freely available on the internet. Due to the ability of PLACET to open Octave environments at execution time, the comfortable script language of Octave can be used for many different purposes. In the integrated simulation framework, Octave controls the program flow and the individual modules, handles processing of the simulation results and the in- and output data. Since Octave also includes libraries of highly optimised numerical functions, the different variants of L-FBs and IP-FBs are implemented in this environment.

3. Controller design

This chapter is concerned with the design of two of the four ground motion mitigation methods used for CLIC, the linac feedback and the IP feedback. While an introduction into these feedback systems was already given in Sec. 1.3.2, this chapter covers more detailed material. After a literature review and a historical overview in Sec. 3.1, conclusions for the orbit feedback controller of CLIC will be drawn. Models for the systems to be controlled will be established and the two feedback algorithm designs will be presented in Sec. 3.2 and Sec. 3.4. To obtain these feedback algorithms the according design procedures will make use of the ground motion models, the models for the beam parameter change due to ground motion and the simulation framework covered in Chap. 2. Since the linac feedback and the IP feedback are not independent of the two other mitigation methods, the characteristic properties of the pre-isolator and the stabilisation system will be taken into account in the design procedure. Since the linac feedback will prove to work very well (see simulation results in Chap. 4), additional cost saving options have been studied in Sec. 3.3. In this work, the possibility of a reduction of the number of necessary correctors is investigated.

Much of the material presented in this chapter will use methods from the field of control system engineering. Control system engineering is mainly concerned about the design and analysis of feedback systems. Real-world systems are abstracted to standardised mathematical models, which are used for the controller design. In App. C, a brief overview of basic control engineering concepts is given. This material is intended for readers which are new to the topic. For a more comprehensive introduction in the paradigms used to describe linear signals and system, the interested reader is referred to Oppenheim, Willsky and Hamid [79]. For a more specific introduction into feedback design methods, the books Frankin, Powell and Emami-naeini [41], Dorf and Bishop [31] and Doyle, Francis and Tannenbaum [33] are recommended.

3.1. Introduction

3.1.1. Classification of accelerator feedback systems

In a modern particle accelerator many different feedback systems are in use. A big distinction can be made between beam-based feedbacks (BB-FB) and feedback loops for auxiliary systems. The most common feedback systems for auxiliary systems are used to control magnet power supplies, RF systems (see Gamp [44] and Schilcher [106]), cryogenics as in Pezzetti [82] and also mechanical stabilisation systems for quadrupoles. Only mechanical stabilisation systems (see Sec. 1.3.2.4) are considered for this thesis, since they are one of the four mitigation methods used in CLIC. Even though the design of theses systems is not covered in this work, their behaviour is included in the all-over luminosity performance simulations in Sec. 4.1. Also the presented controller syntheses

3. Controller design

in this thesis will take these stabilisation system properties into account, to achieve a well harmonised overall performance.

Beam-based feedbacks are control systems in which measurements of beam parameters are used as sensor signals for control systems. Such measurements can be the energy, intensity, transverse tune, phase or position of the beam. For a general introduction into the topic of beam-based feedbacks see Minty and Zimmermann [73]. In the following we will restrict the discussion on beam position feedbacks. Recent development on beam position feedbacks are reviewed in Bulfone [17]. Compact introductions to the theory and application of control engineering for beam position feedback design can be found in Dehler [30], Simrock [122] and Himel [57]. There is not a unique nomenclature for feedback systems falling in this category. We will therefore use in this work the terms defined as follows. Within the group of beam position feedbacks, beam orbit and beam trajectory feedbacks can be distinguished. A trajectory refers to the motion of the centre of a specific bunch $x_i(s)$ along the beam-line length s, where i symbolises the beam bunch index. An orbit on the other hand refers to an average position or expectation value $E\{.\}$ of a collection of bunch centres $\hat{x} = E\{x_i(s)\}$ with $i = 1, 2 \ldots N$. Both feedback algorithms designed in this section (linac feedback and IP feedback) are orbit feedbacks since they act on an average value of the trajectories of individual beam bunches. This is due to bandwidth limitations, since the separation of bunches within a train is only 0.5 ns in CLIC. An example of a trajectory feedback is the intra-train feedback introduced in Sec. 1.3.2.3. Trajectory feedbacks are also used in storage rings to damp single- and multi-bunch instabilities (see Lonza [71]).

3.1.2. Literature review and historical overview

The use of automated orbit feedback systems for storage rings started in the early 80s. Hettel [54] used independent, analogue, single-input single-output (SISO) feedback loops to control the beam orbit at dedicated positions in the Standford Synchrotron Radiation Lightsource (SSRL) ring. The orbit feedback system is referred to in this work as "Steering Servo System". For the vacuum ultraviolet (VUV) ring of the National Synchrotron Light Source (NSLS) a global feedback strategy for the multiple-inputs multiple outputs (MIMO) accelerator system was designed for the first time by Yu [137], which was implemented with analogue technologies. At this time (1989), an automated orbit feedback system was not in use at CERN. However, many beam steering algorithms were implemented and applied as soon as the beam orbit got to large (Brandt et al. [14]) for the Super Proton Synchrotron (SPS), the Large Electron-Positron Collider (LEP) and transfer lines. Major steps forward in the development of orbit feedback systems were reported in Chung [23]. The orbit of the Advanced Photon Source (APS) was controlled with digital technology. Furthermore, an SVD feedback algorithm was used for the first time for this application. Since then, the SVD feedback system became a quasi-standard for storage rings. A prominent example is the orbit feedback system for the Large Hadron Collider (LHC) (Steinhagen [129]). In the case of the LHC the orbit feedback controller is part of several interacting beam-based feedbacks (Steinhagen [128]). Another example for the use of the SVD algorithm is Rowland et al. [104] at the Diamond light source. The reviewed SVD controller implementations are of great interest for this thesis since the designed linac controller also possesses the form of an SVD controller.

3.1. Introduction

For sake of completeness it should be also mentioned, that there are different approaches to beam-based control problems than the classical control techniques mentioned above. Pieck [91] reviews attempts to use methods from artificial intelligence (AI). Bozoki and Friedman [13] e.g. facilitates neural networks to control the beam orbit in a nonlinear storage ring at Brookhaven National Laboratory (BNL). Klein, Westervelt and Luger [62] reports the design of a general purpose control system. A top level controller (expert system) splits up complex tuning and control tasks into smaller subtasks. These tasks are solved using neural networks, fuzzy logic and generic algorithms. A more recent work is Meier et al. [72] in which the structure and weights of a neural network are altered with a reinforcement learning strategy. However, AI methods are not very widespread in the field of orbit control.

For linacs the use of orbit feedback systems started one decade later than for rings, namely in the early 90s. Considerable experience could be gathered from the Stanford Linear Collider (SLC) at the Stanford Linear Accelerator Center (SLAC). Himel [55] introduced the use of the state space formalism in combination with the *linear quadratic Gaussian* (LQG) design procedure for orbit control. Seven independent feedback loops were used to control the beam orbit along the linac. However, no details about the modelling of the subsystems were given. In Barr [10] an adaptive controller (self-tuning regulator) for SLC is presented. A constant gain factor of a local, single degree of freedom controller is altered with the help of a system identification algorithm. Himel [56] also uses adaptive methods for SLC to improve the previous design in Himel [55]. The feedback system in Himel [55] was reported to show overcompensation effects. This was due to the reason that all 7 feedback loops were running independently for different parts of the beam line. A single beam oscillation was measured by all feedback sensors and each loop reacted independently and uncoordinated on the measurement. The problem can be solved if every loop just corrects the errors that are created in the sector controlled by this loop. This is accomplished by passing the measurements of the upstream loop (direction against the beam motion) to the next loop downstream. The upstream data are transformed by the according transfer matrix to the downstream loop, where only the difference between the measurement and the predicted beam position is corrected. It was reported however, that changes of the orbit transfer matrix over time, forced the developers to identify the according matrix with the help of an adaptive learning algorithm. The work of Himel shows the importance of good system knowledge for linac feedback systems and that the quality of this knowledge can be improved by adaptive learning algorithms. This information will be followed up in Chap. 5.

In the end of the 90s the work on the design of a new linear collider and the according orbit and trajectory feedbacks started. Hendrikson et al. [52] summarises the experiences and the current status of the SLAC orbit feedback system as a starting point for future linear colliders. In Hendrickson et al. [53] the status and development of feedback systems for the linear collider projects Next Linear Collider (NLC), TeV-Energy Superconducting Linear Accelerator (TESLA) and Japan Linear Collider (JLC) is reported. Of high interest for this thesis are the statements about performance issues in Hendrikson et al. [53]. It is pointed out that the cascaded, adaptive feedback of Himel [56] performed not as well as initially assumed e.g. for NLC. The causing problem was the wrong assumption that the beam propagation can be described by the multiplication

3. Controller design

of the according local beam transport matrices. However, due to the high acceleration gradients and the according wake-fields and chromatic effects in NLC the beam transport is nonlinear. In other words the beam oscillation depends on its origin. Hendrikson suggests to use a global feedback controller, but mentions practical limitations due to model imperfections. It is also notable, that the orbit feedback design for TESLA was reported to be easier, since super-conducting RF cavities produce weaker wake-fields. As a conclusion, it can be stated that for the orbit control of linear colliders with normal-conducting RF cavities, such as CLIC, global feedback algorithms should be used. For such global algorithms, very good system knowledge is needed.

In the papers of Sery [116] and Hendrickson[51], the focus is shifted from the orbit control along the linac to the IP feedback. A clear distinction between slow ground motion causing mainly emittance growth and faster ground motion responsible for beam-beam offset at the IP is made. Hendrickson suggests the use of a LQG controller for the IP orbit feedback design that can be tuned to specific ground motion spectra. For ground motion spectra with significant cultural noise contributions the luminosity performance was not sufficient. Sery emphasises the importance of ground motion created by auxiliary equipment within the accelerator tunnel and draws the attention to a careful design of such systems. The IP feedback strategy changes with the appearance of the fast trajectory IP feedback FONT. FONT is a very fast bunch-to-bunch feedback system. It uses especially fast electronics to achieve the very short delay times needed. The system developed from FONT1, FONT2 and FONT3 (see Burrows et al.[18]), which are analogue devices, to the latest version FONT4 in digital technology (see Burrows et al. [19]). Summing up, the problem of ground motion for the International Linear Collider (ILC) (see ILC Reference Design Report [90] for more information), which is the successor of the projects NLC, TESLA and JLC, seems to be solved. The reduction of beam-beam jitter due to the FONT4 system in combination with the relaxed tolerances for the orbit feedback system along the linac due to the low wake-fields of the super-conducting RF cavities are reported to result in a strong improvement in terms of luminosity preservation (see White, Walker and Schulte [134] and Sery, Hendrickson and White [117]).

The situation is more difficult in the case of CLIC. The fact that the CLIC beam at the IP is smaller than the ILC beam (about a factor 6 vertically and a factor 14 horizontally) is not even the largest problem. While the ILC beam has a bunch separation time of nominal 369 ns (see ILC Reference Design Report [90]) the CLIC bunches arrive in intervals of only 0.5 ns. In such a regime the FONT system (trajectory feedback controller) cannot work on a bunch-to-bunch basis anymore. However, there are plans to still use FONT3, which is the fastest of all FONT systems since it is still analogue, for the CLIC IP. Since FONT3 is not fast enough to act on each bunch individually it is less efficient in CLIC compared to ILC. An additional pulse-to-pulse feedback controller is necessary (orbit feedback) which will be named in the following IP feedback (IP-FB). The FONT system is kept as a reserve in the current baseline and is therefore not included in the simulations presented in Chap. 4.

For the design of the IP-FB in this thesis, we will follow up the last work done in this area, which was Hendrickson [51]. She stated that the adaptation of the feedback controller to the local ground motion spectrum is an essential task. Caron et al. [20] designs an IP feedback based on a L_2-type controller. To achieve the required performance the

L_2-controller is combined with a mechanical filter systems. Necessary specifications for this mechanical filter system are also given in the same paper.

In Sec. 3.4 an alternative IP-FB design will be presented. This design will be significantly simpler than Caron et al. [20] and less well optimised to the ground motion spectrum. However, due to its simplicity it can be much faster adapted to changing system characteristics. Presented performance predictions will show that the simple controller is operating close to the optimum, which is given by the natural beam jitter.

Also in case of the linac feedback (main linac and beam delivery system) the requirements for CLIC are higher than for ILC. Since CLIC uses normal-conducting RF structures the wake-field effects along the linac are much stronger than for ILC. As a result the beam transport is non-linear with respect to the origin of the beam oscillations. Already some feedback studies on the topic have been performed in the past. Leros and Schulte [69] investigates the emittance preservation in the main linac of CLIC by using a varying number of local, independent feedback loops. In order to reduce the noise amplification and coupling effects between the feedback loops a very low feedback gain is used. An important outcome of the work is that the emittance growth over long times (10 min) is approx. proportional to $1/N_f^2$, where N_f is the number of feedback loops used. Over short time periods (2 s) the feedback system performs well, but only ATL ground motion (see Sec. 2.1) is used as a disturbance, which is not taking into account high-frequency cultural noise. Latina et al. [67] adds to the work done for the main linac a study of the effects in the beam delivery system of CLIC. A strong dependence of the luminosity on the BPM noise is observed and accordingly tight tolerances are inferred. The most recent publication about the orbit feedback of CLIC is Eliasson [35] and the according PhD thesis Eliasson [36]. Eliasson presents a new feedback strategy and several analytic estimates of the effects of different imperfections. He uses detailed system knowledge to control the emittance growth in the main linac of CLIC. Simulations show very good emittance preservation results. The strategy of Eliasson was not followed directly in the current work, since the required system knowledge was very demanding and cannot be measured in practice. However, the presented ideas influenced the current thesis strongly.

3.1.3. Conclusions for CLIC

From the presented literature review, the following important information can be extracted.

- From Hendrikson et al. [53], it is clear that global orbit feedback systems (as Eliasson [35]) perform better than local (as Himel [56]) ones, for ground motion suppression.

- Global feedback systems are already heavily used in storage rings in the form of SVD controller (Steinhagen [129] and Rowland et al. [104]).

- Good system knowledge is essential for global feedback strategies (Hendrickson et al. [53] and Himel [56]).

- For good ground motion mitigation, as many correctors and BPMs as possible should be used (Leros and Schulte [69]).

3. Controller design

- Measurement noise and the according BPM tolerances are a critical cost and performance issue (Latina et al. [67]).

The choice of an orbit feedback algorithm has to balance contrary demands. On the one hand the feedback algorithm should suppress ground motion effects. For this purpose, a global high-gain feedback system with many actors and sensors is preferable. On the other hand the feedback algorithm has to be robust against BPM noise, since the BPM resolution (measurement noise) is a significant cost driver for feedback systems and cannot be chosen arbitrarily small. To make a feedback systems robust against measurement noise, local, low-gain feedback systems are better suited. If a global feedback strategy is chosen, the impact of noise generally gets stronger with an increasing number of used correctors and BPMs, as will be shown in Sec. 3.2.2.2,

We chose to use an SVD controller (global feedback algorithm) in this work, which is already the standard in storage rings. All available actuators and sensors are used, in order to suppress ground motion efficiently. As a result of this choice, most of the design steps presented in Sec. 3.2 are driven by the need to reduce the noise impact on the luminosity performance. The feedback structure (Sec. 3.2.2) as well as the parameter choice (Sec. 3.2.3) will be mainly optimised for noise suppression. The additional necessity of good system knowledge for a global feedback algorithm will be addressed with a system identification algorithm presented in Chap. 5.

3.2. Linac controller

This section is dedicated to the design of the orbit controller for the main linac and the beam delivery system of CLIC. For sake of shortness this feedback controller will be called linac feedback (L-FB), even though the beam delivery system is strictly speaking not an accelerator, since no beam acceleration is performed. The basic problem of beam oscillation damping was already explained in Sec. 1.3.2.1. Section 3.2.1 will formalise this problem in a more quantitative way and introduce a model of beam oscillations in the accelerator. This model will be analysed, to gain important knowledge used in the following design steps. After introducing the system model the structure of the feedback controller will be determined (Sec. 3.2.2). Starting from a basic design, two different strategies for noise reduction will be compared. The chosen structure will be optimised and finalised in Sec. 3.2.3.

3.2.1. Problem statement and system model

3.2.1.1. General system description

As already explained (Sec. 1.3.2.1), the main goal of the linac feedback (L-FB) is to mitigate beam oscillations, which occur mainly due to mechanically misaligned quadrupole magnets. The positions of the magnetic centres of these quadrupoles with respect to a perfect alignment is symbolised by $x[k]$, where k is the time index. The main source of the quadrupole misalignments is ground motion (disturbance $d[k]$). In order to counteract the beam oscillations the beam is steered back onto its reference orbit r_0. The according beam positions are measured therefor with 2122 beam position monitors (BPMs). This BPM sensor data named $y[k]$ already include the measurement noise $n[k]$, which is

3.2. Linac controller

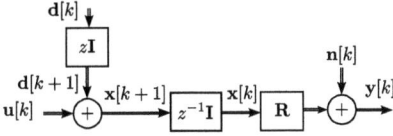

Figure 3.1.: The depicted system models the beam oscillations in the main linac and beam delivery system of CLIC. Shifts of magnetic centres of quadrupoles $x[k+1]$, due to the ground motion $d[k+1]$ and the controller setpoints $u[k]$, are transformed by the unit time shift operator z^{-1} of the $\mathcal{Z}$-transform (see App. C.1) and a matrix multiplication to sensor readings $y[k]$, which include the BPM noise $n[k]$. The ground motion $d[k]$ acts directly on the outputs $y[k]$. Therefore it is multiplied with a factor $z\boldsymbol{I}$, which cancels the time delay of the accelerator.

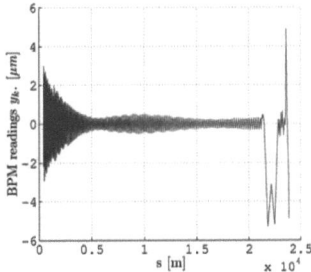

Figure 3.2.: BPM readings due to an unit actuation of 1 μm of the 202nd corrector, where no other imperfections have been applied. The BPM readings show beam oscillations, which are dominated in the main linac by filamentation (beam size increase, see Sec. 1.2) and in the BDS by the lattice design (irregular variations).

assumed to be a white, Gaussian distributed random process throughout this work. In the main linac, one BPM is mounted mechanically to each of the 2010 quadrupoles. Also in the BDS each of the 96 quadrupoles is equipped with a BPM, but there are another intermediate 16 BPMs. The measurements $y[k]$ represent the average beam position of all 312 bunches of a beam train. Since trains are separated from each other by a time delay of 20 ms (see Sec. 1.1) the modelled accelerator system is a discrete system with a sampling rate of $T_d = 20\,ms$.

In the main linac each of the 2010 quadrupoles is equipped with one corrector in horizontal and one in vertical direction. The BPM measurements $y[k]$ are used by a feedback algorithm to calculate the controller setpoint $u[k+1]$ of the next time step. In the BDS all but the last two quadrupoles before the IP, which form the *final doublet*, are equipped with correctors. This corresponds to 94 correctors per direction and per linac. The reason why the last two quadrupoles are not used, will be explained in Sec. 4.2. There is not yet a clear decision on the physical realisation of the correctors. One option is to use the positioning capability of the stabilisation system presented in Sec. 1.3.2.4 to move the quadrupoles mechanically. The second possibility are corrector magnets which are integrated into the quadrupole magnets. Both options shift the effectice magnetic centre of a quadrupole and are, apart from cost issues, basically equivalent. The actuators are assumed in this work to be fast enough to be neglected. In other words the actuators are able to perform a change to the controller setpoint values from $u[k]$ to $u[k+1]$ within the sampling time of 20 ms. However, we will see in Sec. 3.2.3.2 that also much slower actuators do not jeopardise the controller performance.

Since the corrector dynamics are assumed to be much faster than the sampling rate of the L-FB, the change of the sensor readings $y[k]$ due to changes of the correctors

3. Controller design

$u[k]$ can be modelled by a simple matrix multiplication with the *orbit response matrix* R. Its properties will be analysed below. It is important, however, to take into account that controller setpoints applied at time step k to the system are just transferred by R and measured (sampled) one time step later at time index $k+1$ due to the discrete structure of the system. This time delay is the only dynamic element of the accelerator system. The described model is visualised in Fig. 3.1. In this structural view z^{-1} is the unit time shift operator of the $\mathcal{Z}$-transform. For readers not familiar with the $\mathcal{Z}$-transform, an introduction is given in App. C.1. More detailed information can be found in the standard texts of Oppenheim, Schafer and Buck [78] and Franklin, Powell and Workman [42]. Summing up, the accelerator system can be written in the simple form

$$y[k] = Rx[k] + n[k] = Ru[k-1] + Rd[k] + n[k], \qquad (3.1)$$

and its according transfer function

$$Y(z) = Rz^{-1}U(z) + RD(z) + N(z). \qquad (3.2)$$

In Eqs. (3.1) and (3.2) we assume that moveable quadrupoles are used as actuators. Hence, the disturbances and the controller setpoints act on the BPM readings via the the same matrix R.

3.2.1.2. Orbit response matrix R

The orbit response matrix R has 2122 rows and 2104 columns, which corresponds to the number of used BPMs and correctors. A columns of R can be physical interpreted in the following way. The ith column of R corresponds to the BPM measurements (in the unit μm), if the ith corrector is moved by a unit step of $1\,\mu m$ (see Fig. 3.2). A 3D plot of the logarithmic absolute value of the complete R is shown in Fig. 3.3.

Since a dipole kick of a corrector creates only beam oscillations downstream this kick, the BPM measurements upstream are zero. For sake of simplicity only the part of R corresponding to the main linac is investigated first. In the main linac the ith BPM is mechanically mounted to the i+1th quadrupole. As a result R is a triangular matrix if corrector magnets are used. If piezo-movers are used on the other hand R has additional -1 entries in the first diagonal above the main diagonal. These -1 entries originate from the fact that a movement of the i+1th quadrupole by $\Delta u[k, i+1]$ will also move the ith BPM by $\Delta u[k, i+1]$. A beam that arrives at the same position again will result in a BPM reading shifted by $\Delta u[k, i+1]$. This -1 entries are unfavourable for the robustness of the controller. This is due to the fact that with the -1 entries, the controller has the possibility to move the BPMs towards the beam. The initial reference orbit, found by the static alignment methods, is changed gradually. An easy way to work around this problem is the following. When the piezo-actuator moves a quadrupole and the according BPM, the BPM center is changed by the BPM software by the same amount in the opposite direction. As a result the according response matrix has no -1 entries and is triangular. Therefore, the matrix R can be assumed to be triangular and the necessary software BPM center change is omitted in the further discussion.

The part of the orbit response matrix corresponding to only the BDS is not square as the one for the main linac and is therefore not strictly triangular. However the principal

3.2. Linac controller

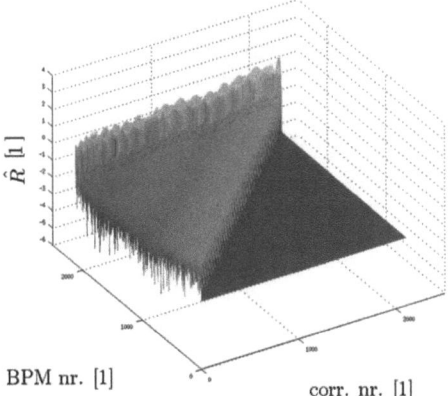

Figure 3.3.: This 3D plot shows $\hat{\boldsymbol{R}} = \log 10\,(|\boldsymbol{R}| + \epsilon)$. The logarithmic scale is used, since otherwise the high beam excursions at the end of the BDS, would not allow observing the typical strong damping by filamentation in the main linac. Due to the logarithmic scale it is necessary to add a small positive contant $\epsilon = 10^{-2}$ to $\boldsymbol{R}$, in order to not map the zero entries due to the quasi-triangular shape onto $-\infty$.

structure is very similar to the one in the main linac. While the form of the matrix entries of the main linac was dominated by filamentation, with the characteristic shrinking of the beam oscillation amplitude, the oscillation amplitudes in the BDS are determined by the lattice design.

The main linac is a beam line consisting of only linear elements (nonlinear acceleration cavities have only a minor effect). Contrary, in the BDS nonlinear elements as sextupoles are present. Beam oscillations originate from an excitation at a certain position with different amplitudes are not just scaled, by a constant but have a nonlinear response behaviour. Therefore, it is questionable if the BDS can be properly approximated with a matrix (description of a linear system). In order to justify the made simplification the following simulations were performed.

As a first step, the response matrix of the BDS (always without BPM noise) was calculated with a beam consisting of 150000 particles. This is three times the normal number of particles, to create simulation results with especially low simulation noise. A corrector step size of $\Delta u = 0.2\,\mu m$ was used for the simulation. The according matrix $\boldsymbol{R_0}$ acts as a reference. As a second step, response matrices with a beam of 50000 particles (nominal number of particles for normal simulation speed) were calculated with different corrector step sizes Δu and compared to R_0. The difference between the matrices is measured by $\Delta R(\Delta u) = ||\boldsymbol{R}(\Delta u) - \boldsymbol{R_0}||_F$, where $||.||_F$ is the Frobenius norm of a matrix, which is defined as $||\boldsymbol{A}||_F = \sqrt{\sum_i \sum_j a_{ij}}$ and corresponds to an quadratic error measure. The quantity $\Delta R(\Delta u)$ is interpreted as a measure for the nonlinearity of the system. Figure 3.4 (left) shows that the matrix deviation can be approximated as

65

3. Controller design

$\Delta R(\Delta u) = a_1 * \Delta u^2$, where $a_1 = 0.0085$.

The questions that remains is what this result means for the deviation of $\boldsymbol{R}$ over longer times for the BDS of CLIC. In reality not only one, but all quadrupoles are moved by ground motion. We relate now the response matrix calculation, where only one quadrupole is misaligned at a time step, to the response matrix of a beam line misaligned by ground motion. For longer time periods such ground motion can be described by the ATL law (see Sec. 2.1), which predicts a motion between two quadrupole of $\sigma_s = \sqrt{ATL}$, where A is the site dependent constant assumed to be $0.5 \times 10^{-6} \mu m^2/s/m$ for the future CLIC site, T is the time difference and L the distance between quadrupoles which is approximated with $27.5\,m$. Since consecutive quadrupole movements are independent of each other (ATL law) and occur at the same time, the effective Δu is estimated to be a factor $\sqrt{96}$ larger than σ_s, where 96 is the number of quadrupoles in the BDS. Combining these approximations, a very simple scaling law $\Delta R(T) = 96 a_1 ATL \approx 10^{-5} T$ can be derived. After ten hours, of ground motion the prediction states a change of $\boldsymbol{R}$ by $R(T) = 10^{-5} \times 3600 \times 10 = 36\%$, which fits well with the data produced by the full-scale simulations performed in PLACET and depicted in Fig. 3.4 (right). In these simulations ATL ground motion was applied for a certain time and $\boldsymbol{R}$ was calculated with $\Delta u = 0.2\,\mu m$. The full-scale simulations also show that for up to one hour of operation not more than 10% of matrix deviation is expected. Since the primary scope of this thesis is the mitigation of ground motion effects up to a few minutes, the modelling of the BDS orbit system as a matrix is justified. For longer time scales long-term mitigation methods will be implemented as explained in Sec. 1.3.3.

3.2.1.3. Control engineering considerations

To summarise from a control engineering point of view, the system to control is a linear, discrete-time system with a sampling time of $20\,ms$. It is a large multi-input, multi-output system (MIMO) with 2104 inputs and 2122 outputs. The 2104 states of the system are the magnetic centres of the quadrupoles $x[k]$. The system is nearly a static one, which can be represented by a simple matrix multiplication. The only dynamic component of the system arises from the fact that an applied corrector change has only an effect at the next time step. This corresponds to a delay element. The corresponding single-input single-output (SISO) analogy has a frequency response representation, that only alters the phase of a signal and multiplies the magnitude with a constant factor. The system has no internal back-coupling and has therefore a finite impulse response (FIR). Since every state can be directly influenced by one actuator the system is clearly controllable. Due to the quasi-triangular structure of the output matrix $\boldsymbol{R}$ with dimension $m \times n$, the system is also observable, since the observation matrix (Kalman)

$$Q_B = \begin{bmatrix} R \\ RA \\ \vdots \\ RA^{n-1} \end{bmatrix} = \begin{bmatrix} R \\ 0 \\ \vdots \\ 0 \end{bmatrix} \in \mathbb{R}^{nm \times n} \quad \text{with} \tag{3.3}$$

$$A = 0 \in \mathbb{R}^{n \times n} \tag{3.4}$$

has full rank n, where $\boldsymbol{A}$ is the system matrix. The observation matrix $\boldsymbol{Q_B}$ has full rank, since already the output matrix $\boldsymbol{R}$ has full rank, because every pair of columns

3.2. Linac controller

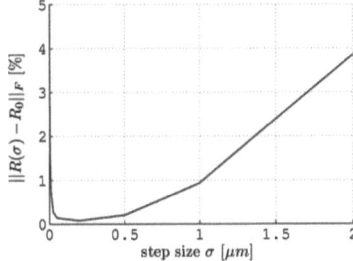 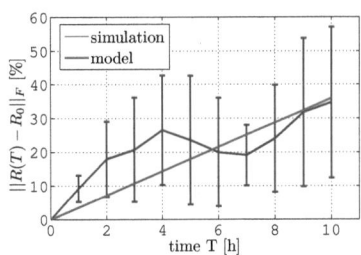

Figure 3.4.: (left) Frobenius norm of the difference between orbit response matrices $R(\Delta u)$ simulated with different step sizes Δu compared to a reference orbit matrix R_0. The system shows a moderately nonlinear behaviour. For $\Delta u < 0.05\,\mu m$ the beam excitations are so small that the measurements are dominated by noise due to the limited number of particles used in the simulation (Schottky noise). (right) Frobenius norm of the change of the orbit response matrix in the BDS of CLIC, due to a misalignment of the accelerator components due to ATL ground motion over 10 hours. The simulation results (blue curve) are an average over 10 seeds. The standard deviations of the simulation results are indicated by error bars. The red curve shows a derived linear scaling law, which fits well with the simulation results for long time periods but underestimates the change of R for shorter simulations. It should be mentioned however that the number of averaged seeds is not very high, since the calculations are computationally expensive.

of R is linear independent (different number of leading zero entries). Therefore, R is of full rank and has more rows than columns. Since the main purpose of the L-FB is to mitigate ground motion effects, a disturbance rejection problem is faced. In the next section two different possibilities to address such a problem will be compared.

3.2.2. Control structure choice

The goal of this section is to find a controller structure which is very well suited to address the special requirements for a linac feedback for CLIC. The term controller structure refers to the fact that in this section no detailed parameter optimisation is performed. Only the type of the feedback controller, which is the basis for a detailed design, is defined. Two complementary feedback structures are compared in this section. One is based on an optimal control formulation of the problem; the other one uses an SVD feedback controller. These methods are tested only in the main linac and not on the BDS for sake of simplicity.

Even though the strategies are quite different, they show strong similarities in their final realisation. This is due to two reasons: they have to fulfil the same requirements and the same accelerator system characteristics are imposed on them. The requirements are fixed by the design choice, made in the previous section, which is the use of one global feedback controller with as many correctors and sensors as possible. This choice implies an inherently good ground motion suppression, but also a strong sensitivity to

3. Controller design

BPM noise. Therefore, the reduction of the influence of the BPM noise will be the main design issue of both concepts.

The nature of the system to be controlled also directs the design of both controller. Due to the huge size of the main linac system (2010 inputs and 2010 outputs) the controller and its design is complicated. It is desireable to reduce the complexity of the controller by limiting the number of design parameters. Such a parameter reduction simplifies the design process, which leads to an controller than can be changed rapidly. To limit the degrees of freedom efficiently, the controller design has to be seen as a two-dimensional problem. The first dimension is the obvious time dimension. But also space, namely the length of the accelerator can be interpreted as a second dimension. The idea that MIMO-Systems have different input/output directions is well-established in control theory (see Skogestad and Postlethwaite [124]), but these directions are normally not interpreted as an additional dimension of the problem, such as time. However, for the accelerator system the interpretation of the input/output directions of the system as an additional dimension is tempting, due to their high number and the different entries of these direction vectors represent the same physical quantity (space). Similarly to the $\mathcal{Z}$-transform in the time-domain the SVD algorithm can be used to find these orthogonal input/output directions in the space-domain. With such an orthogonalisation it is possible to build filters in the space-domain, which separate certain directions from others. In order to reduce the complexity of the design, time-domain and space-domain filters will not be mixed in the following controller designs.

3.2.2.1. H_2-type controller

The design presented in the following was already published in Pfingstner et al. [83]. However, this section will give much more details on the subject. An optimal control problem formulation will be used to design a feedback controller for the main linac. This problem can be written as

$$\min_{C} \mathbb{E}\left\{\boldsymbol{y}[k+1]^T \boldsymbol{y}[k+1]\right\} \tag{3.5}$$

where C is the set of all controller that internally stabilise the system. Such a quadratic optimisation criterion can be solved with the help of the H_2-optimisation theory (see Skogestad and Postlethwaite [124]). For the system structure of the current application the problem simplifies and can be solved by the well studied LQG controller (see Kwakernaak and Sivan [65]). Such a feedback system consists of a state estimator (Kalman-filter) and a state controller (LQR controller), which will be designed in the following. For the further discussion we assume that the reader is familiar with the state space representation of dynamic systems and has basic knowledge about Kalman-filtering. A very brief introduction to Kalman-filtering is given in Appendix C.3. More detailed information can be found in Grewal and Andrews [48] and Kalman [61].

The design of the LQR controller is a trivial problem in our case. This is due to the fact that the system states $\boldsymbol{x}[k]$ can be directly and independently influenced by the actuation $\boldsymbol{u}[k]$; a fact of the assumption that the actuator dynamics is very fast. From Eq. (3.1) it is clear that perfect ground motion suppression would be achieved, if the actuation could be chosen as $\boldsymbol{u}[k] = -\boldsymbol{d}[k+1]$. The overall LQG control problem reduces

3.2. Linac controller

therefore to a ground motion prediction problem, since the control task is trivial. The prediction of ground motion is addressed by the use of Kalman-predictor, which is a slightly modified version of a Kalman-filter.

To predict the ground motion of the next time step, a Kalman-predictor includes a model of this stochastic process. This model has to have to form of a linear dynamic system driven by white noise and hence the ground motion models in Sec. 2.1.2.4 cannot be used directly. To find a dynamic system model that (driven by white noise) describes the ground motion behaviour for all 2010 quadrupoles seems to be an almost impossible task. Every quadrupole is moved by a frequency dependent ground motion PSD, which can be modelled with moderate effort. However, the quadrupoles do not move independently, but are correlated. Considering the high number of inputs, the complex form of the ground motion PSD and the correlation between the quadrupole motions, a suboptimal approach seems to be more promising. In this approach, which is explained in the following, several independent Kalman-predictors are used and the ground motion correlation is neglected due to a lack of according models.

In order to reduce the complexity of the design, as a first step $y[k]$ is multiplied with the pseudo inverse $R^\dagger = (R^T R)^{-1} R^T$, which leads to

$$R^\dagger y[k] = R^\dagger \left(Ru[k-1] + Rd[k] + n[k] \right) = u[k-1] + d[k] + R^\dagger n[k] \qquad (3.6)$$

The pseudo-inverse was used instead of the inverse R^{-1} to show that the technique can also be used for non-square systems. The multiplication of $y[k]$ by $R^\dagger$ decouples the big MIMO system into 2010 SISO systems and, at the same time, aims to reconstruct the states $x[k] = u[k-1] + d[k]$ by inversion. However, the reconstruction is not perfect since also the noise $n[k]$ is multiplied with $R^\dagger$. For each of the 2010 loops an independent Kalman-predictor is now used to reduce the noise level in the ground motion signal.

So far, this strategy neglects the fact that the motion of the individual quadrupoles is correlated in order to simplify the design. However, there is a possibility to still profit from the knowledge of existing correlation. Highly correlated motion corresponds to ground motion with long wave lengths. Such smooth beam-line misalignments are known to degrade beam parameters as emittance and luminosity much less than motion with shorter wave lengths. The idea is now to filter out this smooth, harmless ground motion and reduce in that way also the noise level. Such a filtering is especially effective, since the BPM noise converts (in the main linac) after the multiplication with $R^\dagger$ mainly to a smooth signal. This interesting fact will be clarified in Sec. 3.2.2.2. Without filtering, the signal $R^\dagger n[k]$ would cause very large actuator excursions that would degrade the beam quality and would also exceed the dynamic range of the actuators. The overall feedback system is visualised in Fig. 3.5. In the following we will explain in more detail the three subsystems of the entire control strategy: spatial ground motion filter, noise model and augmented system including a ground motion model.

The ground motion filter has no time dependence and only acts in the spatial dimension. Since it filters out smooth ground motion of long wavelength (small wave number) it is a spatial high pass filter. In the current implementation the vector $R^\dagger y[k]$ is filtered with the help a Hamming-window based, linear-phase, FIR filter created by the Matlab function fir1 from the Signal Processing Toolbox. The cut-off wave length of this high-pass filter was chosen to be 175 m, since these value produced a good trade-off between

3. Controller design

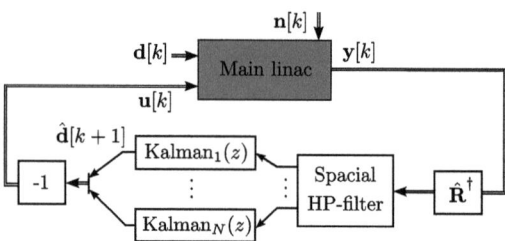

Figure 3.5.: Structure of the H_2-type control algorithm. The multiplication of the BPM measurements $\boldsymbol{y}[k]$ with the pseudo-inverse of the response matrix $\boldsymbol{R}^\dagger$ decouples the inputs. The spatial filter removes smooth ground motion and measurement noise. For each quadrupole one independent Kalman-predictor is used, which produces a prediction $\hat{\boldsymbol{d}}[k+1]$ of the next ground motion vector $\boldsymbol{d}[k+1]$.

noise reduction and signal distortion of the relevant ground motion. Ground motion with shorter wave length degrade the emittance in the main linac strongly, which can be explained by the betatron wavelengths (see Sec. 1.2) of the different main linac sectors (see Schulte [109]). For the creation of the high-pass filter, the function `fir1` needs the parameters N (length (samples) of the filter function) and λ_N (normalised cut-off wave length). Since we chose N to be 500, the filter covers a linac length of $500 \times \Delta L_{BPM}$, where ΔL_{BPM} is the distance between two BPMs here approximated with 10 m. The nth point of a 500 sample long FFT of the BPM readings, corresponds hence to a wave length of $\lambda[n] = 5000\,m/(n-1)$, with $n \in \mathbb{N}^+$. A wave length of 175 m relates thus to $n \approx 30$, which leads to a normalised cut-off wave length of $\lambda_N \approx 30/250 \approx 0.12$, since the function `fir1` normalises the minimal resolvable wave length at data point $N/2$ to 1. This filter is applied to the data with the Octave function `filtfilt`. This function filters the data in forward and reverse direction, which results in a zero-phase filtering, in which the order of the initial filter is doubled. The frequency representation of this filter can be also applied to the 2D PSD of the ground motion (model B). The result is an effective ground motion spectrum (see Fig. 3.6), which is the input for the Kalman-predictor.

Further necessary information for the Kalman-predictor design are the measurement noise properties. A Kalman-predictor requires the noise to be a white, Gaussian distributed stochastic process with zero mean. This is true for the assumed BPM noise $\boldsymbol{n}[k]$, which has a covariance matrix

$$\boldsymbol{N} = \mathbb{E}\left\{\boldsymbol{n}[k]\boldsymbol{n}[k]^T\right\} = \sigma_n^2 \boldsymbol{I}, \qquad (3.7)$$

where $\mathbb{E}\{.\}$ is the expectation value, σ_n is the standard deviation of the BPM noise and $\boldsymbol{I}$ is the identity matrix. The question remains what the properties of the transformed BPM noise at the inputs of the Kalman-predictors are. Since both the multiplication with $\boldsymbol{R}^\dagger$ and the spatial filtering are linear operations and contain no time-dependence, the output noise $\hat{\boldsymbol{n}}[k]$ is also white and Gaussian distributed with zero mean value. If the spatial filter operation is symbolised with f(.), the covariance matrix $\hat{\boldsymbol{N}}$ of $\hat{\boldsymbol{n}}[k]$ can

3.2. Linac controller

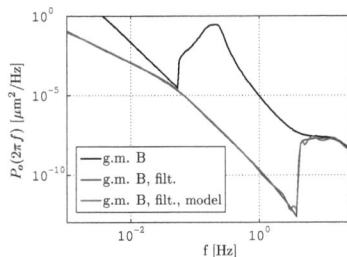

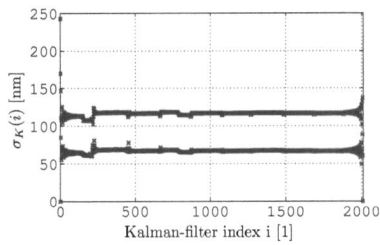

Figure 3.6.: The ground motion spectrum (model B) of the quadrupole movement is plotted (black). Due to the spatial highpass filter the effective spectrum applied to the Kalman-predictor differs (blue). This effective spectrum is modelled by a dynamic system with white input noise (red).

Figure 3.7.: Measurement noise standard deviation $\sigma_K(i)$ for the ith Kalman-predictor. The two line-type formations in the plot ($\approx 70\,\text{nm}$ and $\approx 120\,\text{nm}$) are due to the alternating use of focusing and defocusing quadrupoles in the main linac. Also matching sections can be identified by irregularities in the plot.

be written as

$$\hat{\boldsymbol{N}} = \mathbb{E}\{\hat{\boldsymbol{n}}[k]\hat{\boldsymbol{n}}[k]^T\} = \mathbb{E}\{\text{f}(\boldsymbol{R}^\dagger \boldsymbol{n}[k])\text{f}(\boldsymbol{R}^\dagger \boldsymbol{n}[k])^T\}. \qquad (3.8)$$

Since f(.) is a linear operation,

$$\begin{aligned}\text{f}(\boldsymbol{R}^\dagger \boldsymbol{n}[k]) &= \text{f}(\boldsymbol{r}^\dagger[1]n[1]) + \cdots + \text{f}(\boldsymbol{r}^\dagger[N]n[N]) \\ &= \begin{bmatrix} \text{f}(\boldsymbol{r}^\dagger[1]) & \cdots & \text{f}(\boldsymbol{r}^\dagger[N]) \end{bmatrix} \boldsymbol{n}[k] \equiv \boldsymbol{R}_F^\dagger \boldsymbol{n}[k].\end{aligned} \qquad (3.9)$$

Using the result in Eq. (3.9) in Eq. (3.8) gives the final result

$$\hat{\boldsymbol{N}} = \mathbb{E}\{(\boldsymbol{R}_F^\dagger \boldsymbol{n}[k])(\boldsymbol{R}_F^\dagger \boldsymbol{n}[k])^T\} = \boldsymbol{R}_F^\dagger \mathbb{E}\{\boldsymbol{n}[k]\boldsymbol{n}[k]^T\} \boldsymbol{R}_F^{\dagger\,T} = \sigma_n^2 \boldsymbol{R}_F^\dagger \boldsymbol{R}_F^{\dagger\,T}, \qquad (3.10)$$

where in the last step Eq. (3.7) was used. The noise standard deviations of the individual Kalman-predictors are $\boldsymbol{\sigma}_K = \sqrt{\text{diag}(\hat{\boldsymbol{N}})}$, which are shown in Fig. 3.7.

In order to use a Kalman-predictor, not only the measurement noise $\hat{n}[k,i]$ but also the state noise (in this case $d[k,i]$) has to be a white, Gaussian stochastic process with zero mean value. While the last section showed that this is the case for $\hat{n}[k,i]$, it is not true for $d[k,i]$, which is the ground motion of the ith quadrupole. A common strategy to deal with this problem is to model the state noise with a dynamic system driven by white noise $w[k,i]$. Such a system is commonly called noise shaping filter. The Kalman-predictor is then designed for the combined (augmented) system of the noise shaping filter and the initial system model. To find such a noise shaping filter the relation

$$P_o(\omega) = |H_d(e^{j\omega T_d})|^2 P_i(\omega) \qquad (3.11)$$

can be used, where $P_i(\omega)$ is the Power Spectral Density (PSD) of the input signal to the noise shaping filter with the discrete-time frequency response $H_d(e^{j\omega T_d})$, where

3. Controller design

$T_d = 20$ ms is the sampling time, and $P_o(\omega)$ is the PSD of the output signal of the filter. According to the requirements of the Kalman-predictor, the input signal is chosen to be white noise which corresponds to a constant PSD. For convenience all signals are scaled to $[\mu m^2]$ and therefore $P_i(\omega)$ is chosen to be $P_i(\omega) = 1\,\mu m^2/\text{Hz}$. This corresponds to a variance of the state noise of

$$\sigma_d = \sqrt{\int_{-25\,\text{Hz}}^{25\,\text{Hz}} P_i(\omega) \mathrm{d}f} = \sqrt{1\,(\mu m)^2/\text{Hz}\,(25+25)\,\text{Hz}} = 7.07\,\mu m. \qquad (3.12)$$

Using the assumed $P_i(\omega)$ together with Eq. (3.11), the sought-after noise shaping filter has to have the frequency response

$$|H_d(e^{j\omega T_d})| = \sqrt{P_o(\omega)}. \qquad (3.13)$$

To find such a $H_d(e^{j\omega T_d})$, we assume a rich enough form of $H_d(e^{j\omega T_d}, \boldsymbol{p})$ and vary its free parameters $\boldsymbol{p}$ with a nonlinear optimisation algorithm to match $H_d(e^{j\omega T_d}, \boldsymbol{p})$ to $\sqrt{P_o(\omega)}$. Since $\sqrt{P_o(\omega)}$ ranges over several orders of magnitude, a direct optimisation would not match the two frequency responses well at high frequencies where $P_o(\omega)$ drops quickly. However, this frequency range is still very relevant for the emittance increase. To resolve this problem a target function $J(\boldsymbol{p})$ of the optimisation problem has been formalised in a logarithmic scale. Consequently the optimisation problem can be written as

$$\min_{\boldsymbol{p}} J(\boldsymbol{p}) \quad \text{with} \quad J(\boldsymbol{p}) = \int_{\omega_{min}}^{\omega_{max}} (\log(\sqrt{P_o(\omega)}) - \log(|H_d(e^{j\omega T_d}, \boldsymbol{p})|))^2 \frac{\mathrm{d}\omega}{2\pi}, \qquad (3.14)$$

$$H_d(e^{j\omega T_d}, \boldsymbol{p}) = \frac{k \prod_{i=1}^{3}(z^2 + a_i z + b_i)}{\prod_{i=1}^{2}(z+c_i) \prod_{i=4}^{6}(z^2 + a_i z + b_i)}$$

$$= k \frac{z^6 + \alpha_5 z^5 + \cdots + \alpha_1 z + \alpha_0}{z^8 + \beta_7 z^7 + \cdots + \beta_1 z + \beta_0}\bigg|_{z=e^{j\omega T_d}}, \qquad (3.15)$$

$$\omega_{min} = 2\pi \times 10^{-3}\,\text{Hz} \quad \text{and} \quad \omega_{max} = 2\pi \times 25\,\text{Hz}, \qquad (3.16)$$

where $\boldsymbol{p}$ is the collection of k, a_i, b_i and c_i and $T_d = 20$ ms is the sampling time. Note that since $|H_d(e^{j\omega T_d}, \boldsymbol{p})|$ is an even function with respect to ω, it is enough to formalise the optimisation problem only over positive frequency. As a lower frequency, 0.001 Hz was chosen since components below are not of interest for the dynamic alignment. To ensure that $H_d(e^{j\omega T_d}, \boldsymbol{p})$ is stable and minimum-phase it is sufficient to imposing the constraints

$$|c_i| < 1 \qquad i = 1, 2 \qquad (3.17)$$

$$\begin{bmatrix} -1 & -1 \\ 1 & -1 \end{bmatrix} \begin{bmatrix} a_i \\ b_i \end{bmatrix} < \begin{bmatrix} 1 \\ 1 \end{bmatrix} \qquad i = 1, \ldots, 6 \qquad (3.18)$$

$$|b_i| < 1 \qquad i = 1, \ldots, 6 \qquad (3.19)$$

on the optimisation problem in Eq. (3.14). The conditions in Eqs. 3.18 and 3.19 have been taken from Günther [50]. The optimisation problem above was solved with the help of the Matlab function `fmincon`, which is the general purpose Matlab optimisation function for non-linear, constrained problems. The results are plotted in Fig. 3.6.

3.2. Linac controller

To implement a Kalman-predictor, $H_d(z)$ has to be written in state space form. For that reason, the representation of a transfer function in the canonical normal form (see Föllinger [40]) is used. Since $H_d(z)$ acts directly on the output $y[k]$ (see Fig. 3.1), $zH_d(z)$ has to be used instead of $H_d(z)$. Introducing the state variables of the noise shaping filter $q[k]$, the system can now be written as

$$q[k+1] = \begin{bmatrix} 0 & 1 & & \cdots & 0 \\ 0 & 0 & 1 & & \vdots \\ \vdots & & & \ddots & 1 \\ -\beta_0 & -\beta_1 & -\beta_2 & \cdots & -\beta_7 \end{bmatrix} q[k] + \begin{bmatrix} 0 \\ \vdots \\ 0 \\ 1 \end{bmatrix} w[k]$$

$$\equiv A_d q[k] + b_d w[k] \quad (3.20)$$

$$d[k+1] = \begin{bmatrix} 0 & k\alpha_0 & k\alpha_1 & \cdots & k\alpha_5 & k \end{bmatrix} q[k] \equiv c_d^T q[k]. \quad (3.21)$$

Augmenting this system with a decoupled accelerator system (one quadrupole) gives the final system

$$\hat{x}[k+1] = \begin{bmatrix} q[k+1] \\ x[k+1] \end{bmatrix} = \begin{bmatrix} A_d & 0 \\ c_d^T & 0 \end{bmatrix} \begin{bmatrix} q[k] \\ x[k] \end{bmatrix} + \begin{bmatrix} 0 \\ 1 \end{bmatrix} u[k] + \begin{bmatrix} b_d \\ 0 \end{bmatrix} w[k] \quad (3.22)$$

$$\hat{y}[k] = \begin{bmatrix} 0 & 1 \end{bmatrix} \hat{x}[k] + \hat{n}[k] \quad (3.23)$$

The system in Eqs. (3.22) and (3.23) can easily be checked to be observable by calculating the observability matrix. However, it is not controllable. Obviously the ground motion itself, which is estimated in the noise shaping filter can not be altered. This is not a problem though, since the only state that should be altered, namely $x[k]$, can directly be influenced by the input $u[k]$.

Since the system whose states have be to be estimated, the state noise and the measurement noise are now defined, the classical Kalman-predictor formula can now be applied. Since this is a standard procedure (see Appendix C.3), only the according results are presented in Sec. 3.2.2.3.

3.2.2.2. Weighted SVD controller

A different approach to reduce the effect of BPM noise can be anticipated by analysing the change of the average signal energy of BPM noise, when it is multiplied by an arbitrary matrix. Assuming a random vector $n[k]$, where each of the components $n_i[k]$, $i = 1, \ldots, N$ is a Gaussian distributed random variable with zero mean and a variance σ_n, then the expectation value of the energy of the signal, which is equal to the squared l_2-norm, can be written as

$$\mathbb{E}\{\|n[k]\|_2^2\} = \mathbb{E}\{n[k]^T n\} = \mathbb{E}\left\{\sum_{i=1}^N n_i[k]^2\right\} = \sum_{i=1}^N \mathbb{E}\{n_i[k]^2\} = N\sigma_n^2. \quad (3.24)$$

To analyse the signal energy for $\hat{n}[k] = An[k]$, where A is an arbitrary matrix of dimension $(M_A \times N_A)$, it is useful the write A in its singular value decomposed (SVD) form $A = U_A \Sigma_A V_A^T$. The most important properties of the SVD are summarised in Appendix C.4, while more detailed information can be found in Golub and van Loan [47].

3. Controller design

The reader is only reminded that U_A and V_A are orthonormal matrices and Σ_A is a diagonal matrix containing the singular values $s[i]$ on its main diagonal. Note that the vector $\tilde{n}[k] = V_A^T n[k]$ has the same l_2-norm as n, since V_A is orthonormal. Thus the average signal energy of $\hat{n}[k]$ is

$$\mathbb{E}\left\{||\hat{n}[k]||_2^2\right\} = \mathbb{E}\left\{\hat{n}[k]^T \hat{n}[k]\right\} = \mathbb{E}\left\{\tilde{n}[k]^T \Sigma_A^T U_A^T U_A \Sigma_A \tilde{n}[k]\right\}$$
$$= \mathbb{E}\left\{\sum_{i=1}^{N_A} \tilde{n}_i[k]^2 s[i]^2\right\} = \sigma_n^2 \sum_{i=1}^{N_A} s[i]^2. \quad (3.25)$$

In the third step the orthonormality of U_A was used. The sum over the squared singular values in the last expression of Eq. (3.25) is equivalent to the squared Frobenius norm, which can be also efficiently computed as (see Golub and van Loan [47])

$$||A||_F = \sqrt{\sum_{i=1}^{\min(M_A, N_A)} s[i]^2} = \sqrt{\mathrm{tr}\left(A^T A\right)} = \sqrt{\sum_{i=1}^{M_A} \sum_{j=1}^{N_A} a_{ij}^2} \quad \text{with} \quad A \in \mathbb{R}^{M_A \times N_A}. \quad (3.26)$$

In the following, we exploit this result to design an orbit feedback controller with strong BPM noise demagnification. Similarly as for the H_2-type controller (see Sec. 3.2.2.1), the overall controller is split up into a time-dependent filter and a part that is only dependent on the direction of the measurement vector $y[k] + n[k]$ (spatial filter). We first design the spatial filter and determine the time-dependent filter as a second step. The goal for the design of the spatial filter is to reconstruct the system states $x[k]$ (quadrupole positions) from the beam oscillation signal $y[k]$ and suppress at the same time the BPM noise $n[k]$. Without BPM noise, the states $x[k]$ could be reconstructed perfectly by multiplying the measurement vector with the pseudo-inverse $R^\dagger$. However, since $R^\dagger$ has a large Frobenius norm the unavoidable BPM noise is amplified strongly by this multiplication (see Eq. (3.25)). To resolve this problem, two possibilities to lower the Frobenius norm of the spatial filter are presented in the following.

The first possibility to lower the Frobenius norm of the spatial filter is to use to use less BPMs and correctors. This results in a matrix $R^\dagger$ of lower dimension, which has also a smaller Frobenius norm as can be seen from Eq. (3.26). This method is in agreement with the conclusions from the literature review that the use of more correctors and BPMs increases the BPM noise influence. However, a reduction of the number of correctors and BPMs was excluded in Sec. 3.1.3 in order to mitigate ground motion as efficient as possible.

Another way to reduce the Frobenius norm of the spatial filter has to be found. For that reason, the measurement vector is multiplied not with the full matrix $R^\dagger$, but with a modified matrix $\tilde{R}^\dagger$ in which the magnitudes of the singular values $s[i]$ have been reduced compared to $R^\dagger$. From Eq. (3.25) it is clear that due to this modification the Frobenius norm of $\tilde{R}^\dagger$ is lowered than the one of $R^\dagger$. To determine, which singular values $s[i]$ of $R^\dagger$ have to be lowered, they are plotted in Fig. 3.8. It can be observed that most of the noise amplification comes from $s[i]$ with a high index i and therefore a reduction of these singular values lowers the noise amplification efficiently. It has to be

3.2. Linac controller

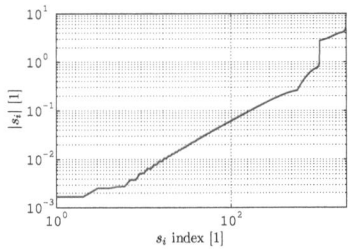

Figure 3.8.: Magnitudes of the singular values $s[i]$ of the matrix $\boldsymbol{R}^\dagger$. Large $s[i]$ contribute stronger to the Frobenius norm of $\boldsymbol{R}^\dagger$ than smaller ones. A large Frobenius norm means a strong average l_2-norm amplification of a signal multiplied with $\boldsymbol{R}^\dagger$.

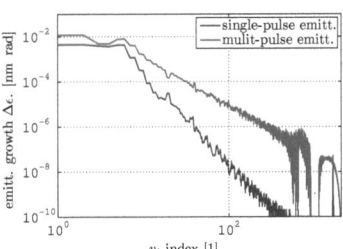

Figure 3.9.: Normalised, projected emittance growth $\Delta\epsilon$ due to the excitation of the beam-line with controller disturbance direction $\boldsymbol{v}[i]$. The $\Delta\epsilon$ for the 17th direction is already a factor ≈ 250 smaller than for the 1st direction (single-pulse emittance).

verified however that this step does not jeopardise the ability of the feedback controller to damp beam oscillations, i.e. to reconstruct the state vector $\boldsymbol{x}[k]$.

To address this point, the significance of the different singular values $s[i]$ and their according BPM and displacement directions $\boldsymbol{u}[i]$ and $\boldsymbol{v}[i]$ (columns of $\boldsymbol{U}$ and $\boldsymbol{V}$) for the emittance growth has to be investigated. For this reason, simulations were performed in which the main linac was misaligned with the displacement direction $\boldsymbol{v}[i]$. The results (shown in Fig. 3.9) make clear that beam oscillations (and hence emittance growth) are very sensitive to the first few $\boldsymbol{v}[i]$, but quite insensitive to the other directions.

Considering the facts that the first few $s[i]$ are capable of mitigating most of the beam oscillations and at the same time only contribute very little to the BPM noise amplification, it is possible to construct a very efficient spatial filter. This is done by not multiplying $\boldsymbol{y}[k]$ with $\tilde{\boldsymbol{R}}^\dagger = \boldsymbol{V}\boldsymbol{F}\boldsymbol{\Sigma}^{-1}\boldsymbol{U}^T$ where $\boldsymbol{F}$ is a diagonal matrix used to scale the singular values in $\boldsymbol{\Sigma}^{-1}$. After testing different configurations, the first 16 diagonal entries of $\boldsymbol{F}$ are chosen to be 1, while all other entries are 0.0001.

As a time-dependent filter we use a simple integrator. Even though this choice seems arbitrary and very simple it has a physical motivation. The ATL law states that ground motion is a random walk in space and time domain. Thus, the change of the ground motion displacement from one to the next time step is independent of the last change. In the case of no BPM noise, the optimal controller action would be to correct the already occurred ground motion change in the next time step, since the change cannot be predicted in advance. This argumentation motivates to use of integrator action in the controller. Note that the ATL law is only valid for long time scales. Thus improvements of the time-dependent filter are possible for shorter time-scales, where predictions of ground motion are possible. The complete feedback structure is visualised in Fig. 3.10.

Even though the controller design in this section was motivated by the need for an efficient BPM noise filter, the final feedback system has the structure of an SVD controller.

3. Controller design

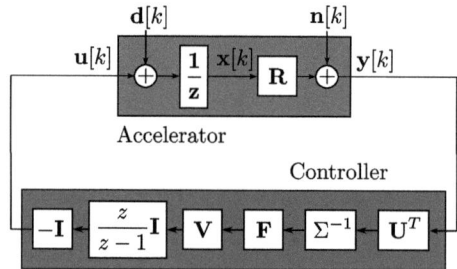

Figure 3.10.: Structure of the all over feedback system. The spatial filter is based on a proper weighting of the singular values of the matrix $R^\dagger$ by the diagonal matrix F, which contains the controller gain factors f_i in its diagonal. The time-dependent parts of the feedback controller are simple integrators with the frequency representation $z/(z-1)$ using the $\mathcal{Z}$-transform.

This can be seen, if we consider that the blocks $z/(z-1)I$ and V can be interchanged in Fig. 3.10. An SVD controller is a *decoupling controller* which separates the global system into N independent single-input single-output loops. This decoupling process is decribed in detail in Sec. 3.2.3.1. For sake of completeness, the according transfer functions are mentioned here, which have for the ith decoupled loop the simple form

$$\hat{Y}(z,i) = z\hat{H}(z,i)\hat{S}(z,i)\hat{P}_v(z,i) - \hat{T}(z,i)\hat{N}(z,i)$$

$$= s[i]\frac{z-1}{z-1+f_i}\hat{P}_v(z,i) - \frac{f_i}{z-1-f_i}\hat{N}(z,i) \qquad (3.27)$$

where $\hat{S}(z,i)$ and $-\hat{T}(z,i)$ are the sensitivity and noise transfer functions as defined in Appendix C.2, $\hat{H}(z,i) = s[i]/z$ is the transfer function of the ith decoupled system channel and f_i is a controller gain factor. The terms $\hat{P}_v(z,i)$ and $\hat{N}(z,i)$ correspond to the spectra of the projected ground motion and noise signals and will be specified in the next section (see Eq. (3.31) and Eq. (2.77)). In Eq. (3.27) $s[i]$ refers to the singular values in $R^\dagger$ and not to the ones in R.

Note that if f_i is equal to 0, the according loop is not closed anymore. The value of the associated integrator in the controller cannot be influenced anymore. Indeed, every transfer function from the input of this integrator to any other signal of the loop is instable. To avoid this problem every f_i is chosen to be different from 0, e.g. $f_i = 10^{-4}$.

3.2.2.3. Choice of the baseline L-FB

The performance of the H_2 controller and the weighted SVD controller are compared in Figs. 3.11 and 3.12. Without feedback control the single- and multi-pulse emittances are very similar, which indicates that the beam jitter is small. When feedback control is used, BPM noise is coupled back and creates mainly beam jitter.

The SVD controller is much better suited for orbit control than the H_2 controller. Even though the time-dependent filters of the SVD controller are not even optimized for

3.2. Linac controller

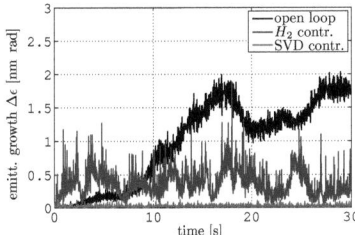

Figure 3.11.: Single-pulse projected emittance growth at the end of the main linac when applying ground motion generated with model B. Even though the H_2 design damps the growth the performance is not sufficient. The influence of BPM noise leads to high-frequent emittance variations.

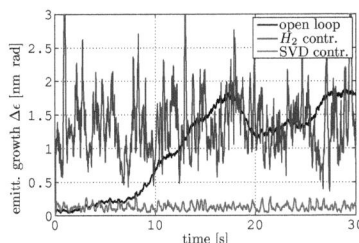

Figure 3.12.: Multi-pulse projected emittance growth at the end of the main linac when applying ground motion generated with model B. The H_2 design even worsens the emittance growth compared to no feedback control (black). The back-coupling of BPM noise creates too much beam jitter.

noise suppression, the emittance preservation is already very good. The reason for this success is the high efficiency of noise suppression of the spatial filter. In the H_2 design, not even the highly optimised Kalman-predictor can compensate for the inadequate noise suppression of the Hamming-window based high-pass filtering. The noise level at the input of the Kalman-predictors is orders of magnitude higher than the signal that should be reconstructed.

SVD controller seem to work especially well on orbit control problems. The reason for that is the nature of the accelerator systems. These systems have a large number of inputs and outputs, which are connected quasi static. Due to the large size of the system, decoupling is advisable to simplify the controller design, which is provided by the SVD controller. Due to this special properties of the accelerator system, the SVD controller decouples the system fully for all frequencies.

3.2.3. Controller optimisation

In this section, the full controller design of the L-FB for the main linac and BDS of CLIC is presented. The L-FB has the form of an SVD controller, which proved to be very well suited for the control of beam oscillations in the last section. The L-FB is optimized to address the special characteristics of the accelerator and the expected ground motion spectra. The design procedure can be split up into three parts. As a first step, the large accelerator system is decoupled into independent channels. Each individual channel controller can be further split up into a time-dependent and a spatial filter. The material in this chapter has been partially published in Pfingstner et al. [89], [86] and [87].

3. Controller design

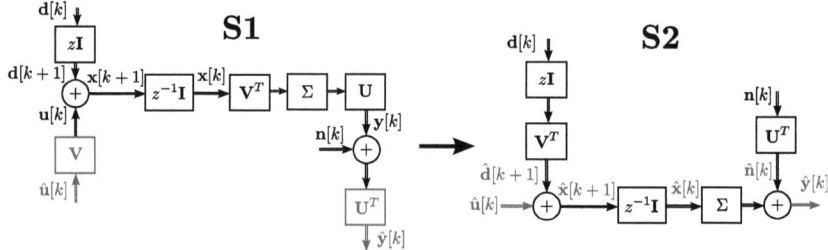

Figure 3.13.: Decoupling of the accelerator system. The accelerator system is pre-multiplied with V and post-multiplied with U^T, to form the virtual system $S1$ (left). The system $S1$ can be transformed to a mathematical equivalent system $S2$ (right), by taking into account that V and zI can be exchanged. Furthermore the identities $VV^T = I$ and $U^TU = I$ are used, since U and V are orthonormal matrices. The decoupling of the in- and outputs of $S2$ becomes apparent, if we consider that Σ is a diagonal matrix.

3.2.3.1. Decoupling

The L-FB is a decoupling controller, which is a common feedback strategy for multi-inputs, multi-output systems (MIMO). The principle of a decoupling controller is to attach to the inputs and/or outputs of the system to be controlled another dynamic system called compensator. The compensator is designed, such that in the combined system each input influences only one output (decoupling). The problem of designing one large controller for a MIMO system can therefore be split up into the design of several single-input, single-output controller (SISO), which is a significant simplification. In general a decoupling controller will not be able to decouple a given system over the whole frequency range equally well. For the given accelerator problem this is possible though, due to the simple structure of the accelerator system which has no internal back coupling.

The special type of decoupling controller used for the L-FB of CLIC is called SVD controller. The decoupling is achieved in this case with the help of the singular value decomposition (SVD) of the orbit response matrix $R = U\Sigma V^T$. The most important properties of the SVD are summarised in Appendix C.4 and more detailed information can be found in Golub and van Loan [47]. Figure 3.13 illustrates the decoupling.

The new system variables of the decoupled system are given by the coordinate transformations $\hat{u}[k] = V^T u[k]$, $\hat{y}[k] = U^T y[k]$ and $\hat{x}[k] = V^T x[k]$. The new in- and outputs do not correspond to individual tripods or BPMs anymore, but to in- and output vector directions given by the columns of U and V. Thus, the change of one input $\hat{u}[k,i]$ varies all tripod actuations $u[k]$ with the pattern $v[i]$, which further changes all beam positions in the BPMs $y[k] - n[k]$ with the pattern $u[i]$. Also the ground motion $d[k+1]$ and the BPM noise $n[k]$ are transformed to $\hat{d}[k+1] = V^T d[k+1]$ and $\hat{n}[k] = U^T n[k]$, to form a compact system representation.

In the following the properties of the decoupled measurement noise vector $\hat{n}[k]$ are

78

3.2. Linac controller

derived. The calculation of the properties of the decoupled ground motion vector $\hat{d}[k]$ is more technical than for the noise $\hat{n}[k]$ and is separately presented in Sec. 2.2.2. Since the elements of $n[k]$ are white, Gaussian, stochastic processes with zero mean, also the elements of $\hat{n}[k] = U^T n[k]$ have the same properties. The covariance matrix of $\hat{n}[k]$ can be calculated by

$$\hat{N} = \mathbb{E}\left\{\hat{n}[k]\hat{n}[k]^T\right\} = \mathbb{E}\left\{(U^T n[k])(U^T n[k])^T\right\} = U^T \mathbb{E}\left\{n[k]n[k]^T\right\} U. \quad (3.28)$$

The matrix $\mathbb{E}\left\{n[k]n[k]^T\right\}$ is a diagonal matrix, since the elements of n_k are uncorrelated. In case all BPMs have the same resolution σ_n, $\mathbb{E}\left\{n[k]n[k]^T\right\} = \sigma_n^2 I$, Eq. (3.28) reduces to

$$\hat{N} = U^T \sigma_n^2 I U = \sigma_n^2 I. \quad (3.29)$$

The transformed noise $\hat{n}[k]$ has in this case exactly the same properties as $n[k]$. If not all BPM have the same resolutions, which is the current baseline (different BPM resolutions for the main linac and the BDS), Eq. (3.29) does not apply and Eq. (3.28) has to be used. In this case the variance vector of the transformed noise $\hat{n}[k]$ is calculated as

$$\sigma_{\hat{n}} = \sqrt{\text{diag}(\hat{N})} = \sqrt{\text{diag}(U^T \mathbb{E}\left\{n[k]n[k]^T\right\} U)}. \quad (3.30)$$

The fact that the elements of $\hat{n}[k]$ are correlated (non-zero off-diagonal elements of $\hat{N}$) is neglected for the controller design, since otherwise the decoupling in individual loop could not be achieved. For the frequency representation of the measurement noise the PSD is used. The PSD of the measurement noise of the ith decoupled channel $\hat{N}(\omega, i)$ is a constant $\hat{N}(\omega, i) = \hat{N}[i]$, since the elements of $\hat{n}[k]$ are white stochastic processes. The value of $\hat{N}[i]$ can be calculated by

$$\sigma_{\hat{n}}^2[i] = 2\int_{f_{min}}^{f_{max}} \hat{N}[i] \mathrm{d}f = 2\hat{N}_i(f_{max} - f_{min})$$

$$\Rightarrow \hat{N}[i] = \frac{\sigma_{\hat{n}[i]}^2}{2(f_{max} - f_{min})}. \quad (3.31)$$

The factor 2 originates from the fact that the integration is only carried out over positive frequencies ($f_{max} > f_{min} > 0$), where the property is used that the PSD is an even function with respect to ω. The boundaries of integration f_{max} and f_{min} are chosen to cover all frequency components of interest.

After the accelerator system has been decoupled, one SISO controller for each of the 2104 independent system channels can be designed. To reduce the number of controller parameters, the choice was made to use the same time-dependent filter $g(z)$ for all channels. Additionally, one constant multiplicative gain factor f_i per channel is added, to be able to account for the different ground motion and noise excitations. The controller of the ith channel has thus the form $g(z)f_i/s[i]$, where the division with the singular value $s[i]$ cancels out the gain factors of the decoupled channels, which have the transfer function $s[i]/z$ (see $S2$ in Fig. 3.13). The complete feedback system is depicted in Fig. 3.14. In this plot the controller is split up into one part that is only time-dependent ($g(z)$) and one part that is only dependent on the direction of the measurement vector

3. Controller design

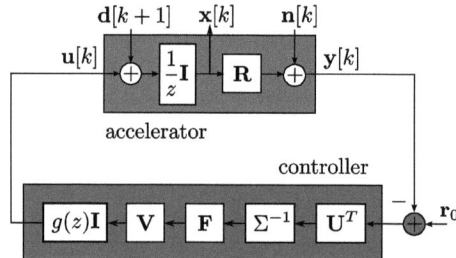

Figure 3.14.: Block diagram of the L-FB. In this structural view, the matrices $g(z)\mathbf{I}$ and $\mathbf{V}$ are exchanged, compared to the explanation in the text, which is mathematically equivalent. The coefficients f_i are collected in the diagonal matrix F. The reference orbit $\mathbf{r}_0$ defines the target orbit and can also be used to create beam bumps along the accelerator.

($\mathbf{VF\Sigma^{-1}U}^T$). The latter part will be referred to in the following as a spatial filter, since the measurement vector represents the spatial distribution of the beam oscillations along the accelerator. The detailed design of the time-dependent and the spatial filter is the topic of the next two sections.

3.2.3.2. Time-dependent filter

The time-dependent filter $g(z)$ is composed of the four elements

$$g(z) = I(z)L(z)P(z)E(z). \tag{3.32}$$

Each of the elements of $g(z)$ represents a discrete-time filter. Since the design of these filters is performed using the $\mathcal{Z}$-transform, the reader is expected to be familiar with this technique. A brief introduction to the $\mathcal{Z}$-transform is given in Appendix C.1, while more detailed information can be found in Oppenheim, Schafer and Buck [78] and Franklin, Powell and Workman [42]. For the following description of the individual elements, basic knowledge about the standard control loop and a technique called loop shaping will be used. A short overview about these topics is given in Appendix C.2.

Integrator $I(z)$: The central element of $g(z)$ is the integrator

$$I(z) = \frac{z}{z-1}. \tag{3.33}$$

The integrating behaviour can be seen, when a signal filtered with $I(z)$ is transformed into the time domain.

$$\mathcal{Z}^{-1}\{Y(z) = I(z)X(z)\} \quad \Rightarrow \quad y[k] = x[k] + y[k-1] \tag{3.34}$$

The use of the $I(z)$ can be physically motivated by the ATL law for ground motion (see Sec. 2.1). This law models the ground motion as a random walk (Brownian motion) in

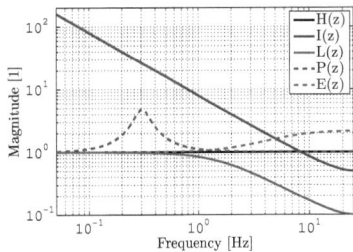
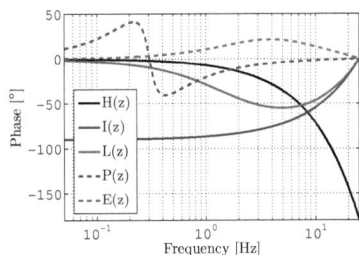

Figure 3.15.: Magnitude (left) and phase (right) of the frequency responses of the elements of $g(z)$ and the normalised accelerator system $a(z) = z^{-1}$. The term normalised refers to the fact that each decoupled system channel has the form $s[i]a(z)$, where $s[i]$ is the ith singular value. The $s[i]$ is canceled by a multiplication with $s[i]^{-1}$ in the controller.

time and space. This means that the increments of the ground motion from one to the next time step are independent. The controller can therefore not make any prediction about the next ground motion disturbance, but can only try to correct the disturbance of the last time step. This corresponds to the reconstruction of the quadrupole position change between the last and the current time step by the multiplication with the spatial filter and adding this correction negatively to the old actuator set points (integrating behaviour). Indeed, if the gain $f_i = 1$ and $g(z) = I(z)$, the controller can correct every ground motion disturbance fully within one time step. This corresponds to a dead-beat feedback, which is optimal for the ground motion suppression, if the disturbance would be only ATL motion and there would be no BPM noise.

The frequency response of $I(z)$ is depicted in Fig. 3.15 (blue curve). The high gain at low frequencies ensures good ground motion suppression, while the low gain at high frequencies leads to a signal amplification in this frequency range. From Eq. (3.27) it is clear that the gain factor f_i cannot be chosen larger than 2 in case $g(z) = I(z)$, since otherwise the control loop gets instable which can be shown with the Nyquist criterion (see Appendix C.2.2 for more explanations). Even though $I(z)$ suppresses ground motion effects already very well, the BPM noise is fully fed back into the accelerator system, which degrades the luminosity considerable (see Sec. 4.2.1).

Low pass $L(z)$: To improve the noise behaviour of the controller the low pass

$$L(z) = \frac{z\left(1 - e^{-\frac{T_d}{T_1}}\right)}{z - e^{-\frac{T_d}{T_1}}} \quad \text{with} \quad (3.35)$$

$$T_d = 0.02\,\text{s} \quad \text{and} \quad T_1 = 0.1\,\text{s},$$

is added to $g(z)$, where T_d is the sampling time. The choice of the T_1 was made due to a simplified luminosity estimate presented in Sec. 3.4 according to model Eq. (2.74).

3. Controller design

As can be seen from the red curve in Fig. 3.15, $L(z)$ demagnifies signals above its cutoff frequency of about 1.4 Hz.

The design of $L(z)$ has been carried out using the Z-transform. The operation Z-transform should not be confused with the $\mathcal{Z}$-transform described in Appendix C.1. The Z-transform is a procedure to calculate the transfer function $L(z)$ of a discrete system, which consists of a continuous system $L(s)$ (Laplace-transform representation) that has at its input a zero-order hold element and is sampled at its output. This configuration is typical for continuous systems, which are controlled by a discrete-time controller. The according discrete time system transfer function $L(z)$ can be calculated by

$$L(z) = (1 - z^{-1})Z\left\{\frac{L(s)}{s}\right\}, \qquad (3.36)$$

where $Z\{.\}$ corresponds to a series of the following sequential operations. The argument (Laplace-transform of a system devided by s) is transformed into the time domain by applying the inverse Laplace-transform $\mathcal{L}^{-1}\{.\}$. The resulting time function is sampled at the points $t = kT_d$. Finally, this series representing a time-discrete signal is $\mathcal{Z}$-transformed (see Gausch, Hofer and Schlacher [46] and Günther [50], where the latter text uses instead of Z-transform the term ζ-transform). For the current application a first-order low pass $L(s)$ forms the basis for the calculation of

$$\hat{L}(z) = (1 - z^{-1})Z\left\{\frac{1}{s(1 + sT_1)}\right\}, \qquad (3.37)$$

where T_1 is the time constant of the low pass. The evaluation of Eq. (3.37) can be done directly or via a table lookup (see Günther [50]). The resulting $\hat{L}(z)$ has a quite large phase shift for high frequencies, which is disadvantageous for the stability properties. Even though this phase change is indispensable for the sampled continuous system, it can be avoided in the discrete realisation by choosing $L(z) = z\hat{L}(z)$, which gives the final expression in Eq. (3.35).

Peak $P(z)$: The element $P(z)$ has to be introduced to address an issue arising from the special topology of CLIC. The final doublet (FD) quadrupoles are stabilised by the pre-isolator, while for the other quadrupoles of the main linac and BDS the quadrupole stabilisation system is used (see Sec. 1.3.2). Since the frequency responses of both systems are different (Fig. 1.8 (right)), even fully correlated ground motion would lead to an offset between the FD and the rest of the quadrupoles. By neglecting the pre-isolator tilt and correlation, a simplified model for the PSD of the differential motion $P_{FD}(\omega)$ can be found. Using the modelling framework presented in Sec. 2.2.1, $P_{FD}(\omega)$ can be written as

$$P_{FD}(\omega) = |G_{PRE}(j\omega) - G_{STAB}(j\omega)|^2 \int_{-\infty}^{+\infty} P_{B10}(\omega, k)\frac{\mathrm{d}k}{2\pi}, \qquad (3.38)$$

where $G_{PRE}(j\omega)$ and $G_{STAB}(j\omega)$ are the frequency responses of the pre-isolator (point-like model) and the quadrupole stabilisation. The term $P_{B10}(\omega, k)$ symbolises the two-dimensional PSD of the ground motion model B10. As can be seen in Fig. 3.17, the offset is especially large around 0.3 Hz if the quadrupole stabilisation version 1 is used

3.2. Linac controller

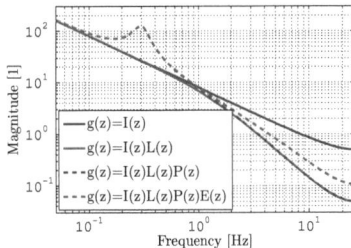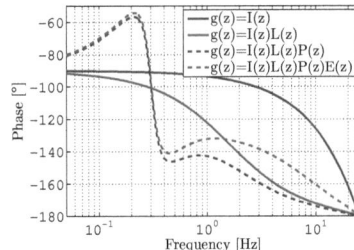

Figure 3.16.: Magnitude (left) and phase (right) of the frequency response of the open loop transfer function $(s[i]z^{-1})(g(z)f_i/s[i])$. The influence of the elements of $g(z)$ is investigated by sequentially adding them to the controller. Note that the depicted frequency responses are symmetric around 25 Hz and periodic around 50 Hz, since they are discrete-time. This means that frequencies with $n \times 50$ Hz, where $n \in \mathbb{N}$, are treated the same way by the controller.

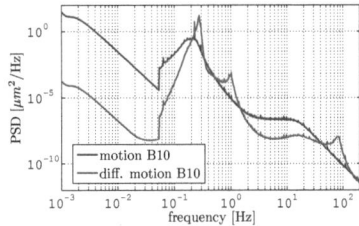

Figure 3.17.: The blue curve shows the PSD of the absolute motion of the FD (point-like model). The PSD of the difference motion between the FD and rest of the accelerator $P_{FD}(\omega)$ is plotted in red, where the quadrupole stabilisation version 1 was used for all, but the FD quadrupoles.

(see Sec. 1.3.2.4), since the ground motion has strong components in this frequency range (microseismic peak).

The quadrupole offset can cause a beam offset in the FD, which results in secondary, luminosity diluting effects. A model for the main effect is given in Sec. 2.3.2. An experiment in form of a simulation was conducted, to verify that the offset of the FD quadrupoles with respect to the other quadrupoles is not in a principle problem for the luminosity. The FD was offset compared to the rest of the accelerator and the L-FB was switched on. No ground motion or other imperfections were applied. After only a few time steps the L-FB could recover the luminosity fully. This proved that the observed luminosity decrease was not due to the kink in the beam trajectory when it was steered into the FD. The beam could simply not be steered fast enough into the FD by the L-FB, due the fast changing offset. Following this observations, it seemed to be advantageous to improve the controller performance in the frequency range around 0.3 Hz with the

83

3. Controller design

element

$$P(z) = \frac{(1-n_1)(1-n_2)}{(1-z_1)(1-z_2)} \frac{(z-z_1)(z-z_2)}{(z-n_1)(z-n_2)} \quad \text{with} \tag{3.39}$$

$$z_{1,2} = e^{(-1.43 \pm 2\pi i 0.2)T_d} \quad \text{and} \quad n_{1,2} = e^{(-0.3 \pm 2\pi i 0.3)T_d}. \tag{3.40}$$

The magnitude of the frequency response of $P(z)$ has a peak around 0.3 Hz. This peak is designed to be as high as possible. At the same time, the peak should influence high frequencies as little as possible in order to not amplify noise strongly (see dashed blue curve in Fig. 3.15). The basis for the design of such a transfer function is a low pass formed by the poles of $P(z)$. This low pass is designed to have a considerable overshoot before the cutoff frequency, which is achieved by a conjugate complex pair of poles. The cutoff frequency is adjusted such that the peak of the overshoot is located at 0.3 Hz. To not influence high frequencies, the zeros of $P(z)$ cancel out the demagnification of the low pass at high frequencies and also enhance the peak at 0.3 Hz.

The location of the poles and zeros was hand-tuned until the desired effect was achieved. The poles and zeros are written as powers of e in Eq. (3.40). This form makes it easy to design a stable and minimum phase $P(z)$, by keeping the real part of the argument negative.

Phase lifting element $E(z)$: The combination of the elements $I(z)$, $L(z)$ and $P(z)$ leads to a relatively small phase margin $\Delta\phi_c$ of the open loop frequency response (see Appendix C.2.2 for a definition). This is disadvantageous for the stability properties of the system, as is explained in Appendix C.2.2. To improve the situation a phase lifting element

$$E(z) = \frac{(1-n_3)(z-z_3)}{(1-z_3)(z-n_3)} \quad \text{with} \tag{3.41}$$

$$z_3 = e^{-17T_d} \quad \text{und} \quad n_3 = e^{-38T_d},$$

is added to $g(z)$. The pole and the zero are adjusted, such that $\Delta\phi_c$ is increased to 36.3° as can be seen in Fig 3.16 (right). Note that this $\Delta\phi_c$ corresponds to a $f_i = 1$. As will turn out in the design of the spatial filter, the gain factor f_i are always chosen smaller than 1, which further increases $\Delta\phi_c$. The noise is only slightly amplified by $E(z)$ at high frequencies as can be seen in Fig. 3.16 (left).

The complete time-dependent filter $g(z) = I(z)L(z)P(z)E(z)$ results in the open loop frequency response in Fig. 3.16 and the sensitivity and noise transfer functions $\hat{S}(z)$ and $-\hat{T}(z)$ in Fig. 3.18.

Comparison to the q-transform design: The controller transfer function $g(z)$ found in this section was designed by shaping the time-discrete, open loop frequency response directly via proper compensator elements. Another commonly utilised design method uses the *q-transform* to transform the time-discrete system into a quasi-continuous one. The design can then be performed in the q-domain by using the well known continuous loop shaping method. For sake of completeness, also the q-transfer function $g(q)$ of the

3.2. Linac controller

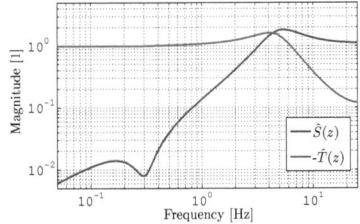 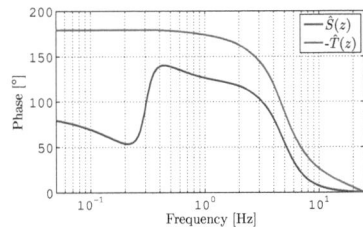

Figure 3.18.: Magnitude (left) and phase (right) of the decoupled sensitivity function and noise frequency response $\hat{S}(e^{j\omega T_d})$ and $-\hat{T}(e^{i\omega T_d})$ for $f_i = 1$. The ground motion is strongly demagnified at low frequencies by $\hat{S}(z)$. Notice the stronger demagnification around 0.3 Hz, due to the element $P(z)$. At higher frequencies, ground motion is only modestly amplified around 5-6 Hz. The BPM noise is directly transmitted to the beam by $\hat{N}(e^{j\omega T_d})$ at low frequencies, but is demagnified at higher frequencies.

already designed controller $g(z)$ is calculated, plotted and analysed below. The function $g(q)$ can be calculated by applying the bilinear transform

$$z = \frac{1 + q\frac{T_d}{2}}{1 - q\frac{T_d}{2}} \qquad (3.42)$$

to $g(z)$ in Eq. (3.32) (variable substitution), which results in

$$g(q) = I(q)L(q)E(q)P(q) \qquad (3.43)$$

with

$$I(q) = \frac{q\frac{T_d}{2} + 1}{qT_d} \qquad (3.44)$$

$$L(q) = \frac{q\frac{T_d}{2}\left(1 - e^{-\frac{T_d}{T_1}}\right) + \left(1 - e^{-\frac{T_d}{T_1}}\right)}{q\frac{T_d}{2}\left(1 + e^{-\frac{T_d}{T_1}}\right) + \left(1 - e^{-\frac{T_d}{T_1}}\right)} \qquad (3.45)$$

$$P(q) = \frac{(1 - n_1)(1 - n_2)}{(1 - z_1)(1 - z_2)} \left[\frac{q\frac{T_d}{2}(1 + z_1) + (1 - z_1)}{q\frac{T_d}{2}(1 + n_1) + (1 - n_1)}\right] \left[\frac{q\frac{T_d}{2}(1 + z_2) + (1 - z_2)}{q\frac{T_d}{2}(1 + n_2) + (1 - n_2)}\right] \qquad (3.46)$$

$$E(q) = \frac{(1 - n_3)}{(1 - z_3)} \left[\frac{q\frac{T_d}{2}(1 + z_3) + (1 - z_3)}{q\frac{T_d}{2}(1 + n_3) + (1 - n_3)}\right]. \qquad (3.47)$$

The element $I(q)$ is a continuous *proportional integral compensator* (PI), $P(q)$ a continuous lead-lag element and $E(q)$ a continuous lead-element. The q-frequency responses to these q-transfer functions are given by evaluating the complex variable q at $j\Omega$. To plot the open loop q-frequency response in Fig. 3.19, also the time-discrete transfer function of a decoupled accelerator channel $\hat{H}(z, i) = s[i]/z$, where $s[i]$ is the singular value of

3. Controller design

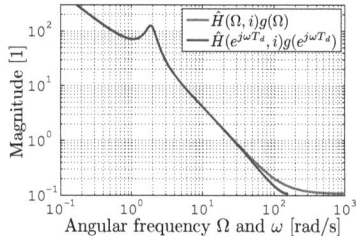 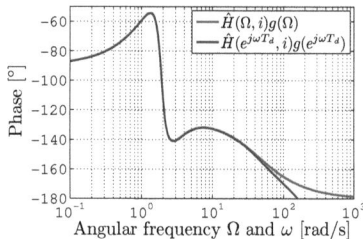

Figure 3.19.: Magnitude (left) and phase (right) of the decoupled open loop q-frequency response $\hat{H}(\Omega,i)g(\Omega)$ and the discrete-time frequency response $\hat{H}(e^{j\omega T_d},i)g(e^{j\omega T_d})$, where $s[i]$ (singular value of the decoupled channel) is assumed to be 1. The two frequency responses are very similar for $\omega T_d \ll 1$. For higher angular frequencies the two functions differ, which is a well known property of the q-transform.

the decoupled channel, has to be q-transformed which gives

$$\hat{H}(q,i) = \frac{-qs[i]\frac{T_d}{2} + s[i]}{q\frac{T_d}{2} + 1}. \qquad (3.48)$$

Since the bilinear transform is only distorting the frequency response along the angular frequency axis, the phase margin $\Delta\phi_c$ does not change and all statements regarding the stability, made in the section about the compensator $E(z)$, are still valid. Additionally, the characteristic values of a step response, the rise time T_r (time the tangent on the step response at 50 % of the step size needs to rise from 10 % to 90 % of the step size) and the overshooting width u_o can be estimated from the q-frequency response with the empirical formula

$$\Omega_c T_r \approx 1.2 \quad \text{and} \qquad (3.49)$$
$$\Delta\phi_c[°] + u_o[\%] \approx 70, \qquad (3.50)$$

which are taken from Gausch, Hofer and Schlacher [46]. Applying the values $\Omega_c = 27.99\,\text{rad/s}$ and $\Delta\phi_c = 36.3°$ to these formula, the characteristic values of a setpoint step $T_r \approx 43\,\text{ms}$ and $u_o \approx 34\,\%$ are found. It should be mentioned however that the controller is optimised for disturbance rejection and not for setpoint following.

3.2.3.3. Spatial filter

In this section the design of the diagonal elements f_i of the diagonal scaling matrix $\boldsymbol{F}$ is presented. In case the scaling would not be used, or equivalently $\boldsymbol{F} = \boldsymbol{I}$, the spatial filter $\boldsymbol{VF\Sigma^{-1}U^T}$ would be simply the pseudo-inverse of the matrix $\boldsymbol{R}$. The technique of scaling the singular values for the design of orbit feedback controller was already used before in literature. For example, Rowland et al. [104] uses a Tikhonov regularisation (see Neumaier [77] for details to this method) to decrease the conditioning number (ratio of the largest to the smallest singular value) for $\boldsymbol{R}$ to improve the numerical properties of

3.2. Linac controller

the calculation of the pseudo-inverse. The scaling presented in this section goes beyond an improvement of the matrix inversion properties. The f_i are chosen to minimise the output signal $\hat{y}[k,i]$ of each individual loop, by using models of the disturbance (ground motion) and measurement noise. Since the spectra of these input signals vary from channel to channel, it is necessary to optimise each f_i individual. The PSD of the output signal $\hat{y}[k,i]$ of the ith control loop can be calculated as

$$\hat{Y}(\omega,i) = \left|z\hat{H}(e^{j\omega T_d},i)\hat{S}(e^{j\omega T_d},i)\right|^2 \hat{P}_v(\omega,i) + \left|-\hat{T}(e^{j\omega T_d},i)\right|^2 \hat{N}(\omega,i), \qquad (3.51)$$

where the PSD of the virtual ground motion excitation and the measurement noise are symbolised with $\hat{P}_v(\omega,i)$ and $\hat{N}(\omega,i)$. The terms $\hat{S}(e^{j\omega T_d},i)$ and $-\hat{T}(e^{j\omega T_d},i)$ are the sensitivity function and noise frequency respons of the ith decoupled closed loop. Their $\mathcal{Z}$-transforms are given by

$$\hat{S}(z,i) = \frac{1}{1+\hat{H}(z,i)\hat{C}(z,i)} \quad \text{and} \quad \hat{T}(z,i) = \frac{\hat{H}(z,i)\hat{C}(z,i)}{1+\hat{H}(z,i)\hat{C}(z,i)} \quad \text{with} \qquad (3.52)$$

$$\hat{H}(z,i) = \frac{s[i]}{z} \quad \text{and} \quad \hat{C}(z,i) = g(z)\frac{f_i}{s[i]}, \qquad (3.53)$$

where $\hat{H}(z,i)$ and $\hat{C}(z,i)$ are the transfer functions of a decoupled accelerator channel and the according controller. For a derivation of the expressions for $\hat{S}(z,i)$ and $\hat{T}(z,i)$, please refer to Appendix C.2. The factor z in Eq. (3.52) originates from the way the ground motion is modelled, which can be seen in Fig. 3.13. To calculate $\hat{Y}(\omega,i)$ the PSDs of the ground motion $\hat{P}_v(\omega,i)$ and the BPM noise $\hat{N}(\omega,i)$ have to be known. The derivation of these spectra was carried out in Eq. (3.30) and Eq. (3.31) for $\hat{N}(\omega,i)$ and in Sec. 2.2.4 and Sec. 2.2.2 for $\hat{P}_v(\omega,i)$.

Even though the expressions in Sec. 2.2.2 offer a comfortable way to calculate $\hat{P}_v(\omega,i)$ in a closed form, some numerical problems were encountered in the practical application. These problems arise from inaccuracies of the response matrix $\boldsymbol{R}$, which is used for the calculation of $\hat{P}_v(\omega,i)$. When calculating the matrix $\boldsymbol{R}$ with the beam tracking code PLACET, only a strongly reduced number of particles, compared to the real CLIC beam, can be used in the simulations. This limited number of particles leads to a coarser beam distribution as in reality, which results in artificial beam fluctuations measured by the BPMs (Schottky noise). These artificial fluctuations create errors in $\boldsymbol{R}$. For the presented results a different way for the calculation of $\hat{P}_v(\omega,i)$ was chosen. Via simulations with the assumed ground motion model, the output signals $\hat{y}[k,i]$ were created, stored and Fourier-transformed. The L-FB was turned off, but the quadrupole stabilisation was used. Since these signals are random processes, the simulations had to be carried out for several different seeds of the ground motion random generator and the resulting signal spectra were averaged.

Since all involved transfer functions and signals are now defined, the optimal value for f_i can be found by minimising the power of each signal $\hat{y}[k,i]$. From Parseval's theorem it is known that this minimisation problem is equivalent to the minimisation of the L_2-norm of the Fourier-transform of $\hat{y}[k,i]$, which corresponds in this case to the

3. Controller design

PSD $\hat{Y}(\omega, i)$. Hence, the f_i can be found by

$$\min_{f_i} ||\hat{y}[k,i]||_{pow} = \min_{f_i} \int_{\omega=-\infty}^{+\infty} \hat{Y}(\omega, i, f_i) d\omega \quad \forall i : i = 1, 2, \ldots 2104 \quad (3.54)$$

Since there is no closed expression for $\hat{P}_v(\omega, i)$ available, the integral in Eq. (3.54) was calculated numerically and the minimisation problem was solved with a parameter scan of f_i. The integration was only carried out over positive frequencies, since $\hat{Y}(\omega, i, f_i)$ is symmetric and the negative frequencies can be taken into account by a multiplication with a factor 2. Furthermore, the integration is not carried out from 0 to $+\infty$, but only over a large enough frequencies range to cover all significant frequency components.

The luminosity preservation performance of the L-FB can be further improved, if also the action of the IP-FB is taken into account. As explained in Sec. 1.2, the beam size growth at the IP but also the beam-beam offset create luminosity loss. The L-FB influences both effects with the same frequency response, since the loop output $\hat{y}[k, i]$ corresponds to a beam motion with the spatial shape $u[i]$, which creates beam size growth as well as beam-beam offset. The beam-beam offset however, is additionally multiplied by the disturbance rejection frequency response of the IP-FB $H_{IP}(e^{j\omega T_d}) S_{IP,2}(e^{j\omega T_d})$, where $H_{IP}(e^{j\omega T_d}) = z^{-1}$ and $S_{IP,2}(e^{j\omega T_d})$ is given in Eq. (3.73). Hence, for the beam growth the PSD $\hat{Y}(\omega, i)$ is relevant, while the beam-beam offset is described by $\left|z^{-1} S_{IP,2}(e^{j\omega T_d})\right|^2 \hat{Y}_v(\omega, i)$. To combine both signals into one joined cost function, the relevance of the signals for the according luminosity loss has to be determined for each decoupled loop.

For this reason, a simulation was performed, which is very similar to the one described in Sec. 2.2.3, where a model for the luminosity loss due to ground motion was created. While in Sec. 2.2.3, the elements of the accelerator were misaligned with sine and cosine waves, the input vectors of the decoupled system channels $v[i]$ are now used to misalign the beam line. The vectors $v[i]$ only cover either the electron or the positron part of the accelerator. To evaluate the effect on the luminosity correctly, both parts have to be misaligned however. Therefore, new input directions $\tilde{v}[i]$ have to be defined. These $\tilde{v}[i]$ can be found by performing an SVD on an orbit response matrix $\tilde{R}$, which covers both accelerator arms. To connect both arms in $\tilde{R}$ a virtual BPM at the IP was added, which leads to the definition

$$\tilde{R} = \left[\begin{bmatrix} R_{IP} \\ \mathbf{0}^{m \times n} \end{bmatrix} \begin{bmatrix} \mathbf{0}^{m \times n} \\ R_{IP,F} \end{bmatrix} \right] \in \mathbb{R}^{(2m+1) \times n} \quad \text{with} \quad (3.55)$$

$$R_{IP} = \begin{bmatrix} R \\ r_{IP}^T \end{bmatrix} \in \mathbb{R}^{(m+1) \times n} \quad \text{and} \quad (3.56)$$

$$R_{IP,F}[i,j] = R_{IP}[m-i+2, n-j+1] \in \mathbb{R}^{(m+1) \times n}. \quad (3.57)$$

In Eq. (3.56), the vector r_{IP} describes the influence of the quadrupole offsets on the beam offset at the virtually added BPM at the IP. The new matrix for one accelerator arm R_{IP} can be combined with its flipped version $R_{IP,F}$ to form the response matrix $\tilde{R}$ for the complete accelerator complex. For the combination the fact is used that both accelerator parts are symmetric. The flipping of $\tilde{R}$ (defined in Eq. (3.57)) is necessary to allow a sequential ordering of the BPMs and quadrupoles along the accelerator. When

3.2. Linac controller

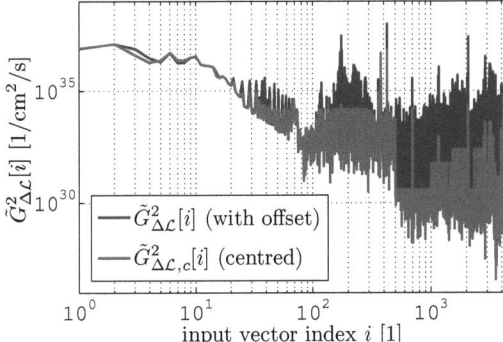

Figure 3.20.: Normalised luminosity loss due to a misalignment of the accelerator with the vectors $\tilde{v}[j]$, which corresponds to the sensitivity functions $\tilde{G}^2_{\Delta\mathcal{L}}[i]$ discussed in Sec. 2.2.4. The red curve correspond to the normalised luminosity loss $\tilde{G}^2_{\Delta\mathcal{L},c}[i]$, if the beams where shifted against each other until the max. luminosity value is reached (centred). The blue curve belong to the normalised luminosity loss of the non-shifted beams $\tilde{G}^2_{\Delta\mathcal{L}}[i] = \tilde{G}^2_{\Delta\mathcal{L},c}[i] + \tilde{G}^2_{\Delta\mathcal{L},o}[i]$, where the index o stands for offset.

the matrix $\tilde{R}$ is decomposed by the SVD, it turns out that the 4208 input directions $\tilde{v}[j]$ are closely related to the 2104 input directions of one accelerator arm $v[i]$. A pair of two $\tilde{v}[j]$ can always be created out of one corresponding $v[i]$ by

$$v[i] \in \mathbb{R}^n \quad \rightarrow \quad \tilde{v}[2i-1] = \begin{bmatrix} v[i] \\ v_F[i] \end{bmatrix} \quad \text{and} \quad \tilde{v}[2i] = \begin{bmatrix} v[i] \\ -v_F[i] \end{bmatrix} \quad (3.58)$$

where again $v_F[i]$ is the flipped version of the vector $v[i]$. The vectors $\tilde{v}[2i-1]$ and $\tilde{v}[2i]$ form a pair of symmetric and anti-symmetric misalignments, similar to the cosine and sine waves in Sec. 2.2.3. If two beams are tracked through the accelerator misaligned with $\tilde{v}[2i-1]$, they collide with no beam-beam offset, since the electron and positron beam is influenced exactly the same way. The according luminosity loss is solely due to the beam size growth. To be able to separate the effects of beam-beam offset and beam size growth, the beams are shifted against each other until the maximum luminosity value is observed. If the accelerator is misaligned with $\tilde{v}[2i]$, the beam size growth as well as the beam-beam offset contributes to the luminosity loss. Note that the luminosity loss only due to beam size growth does not need to be the same for $\tilde{v}[2i-1]$ and $\tilde{v}[2i]$. This is due to the fact that the 2-dimensional beam histogram in vertical and horizontal direction is not necessarily symmetric, due to nonlinear effects in the BDS. The simulations had to be carried out for different excitation amplitudes to stay in a proper luminosity regime (see Sec. 2.2.3 for more explanations). The final results are the sensitivity functions $\tilde{G}_{\Delta\mathcal{L}}[i]$ and $\tilde{G}_{\Delta\mathcal{L},c}[i]$, which are depicted in Fig. 3.20.

In the practical realisation two identical, independent L-FBs are used for the electron and positron part of the accelerator. These controller process $\tilde{v}[2i-1]$ and $\tilde{v}[2i]$ the same

3. Controller design

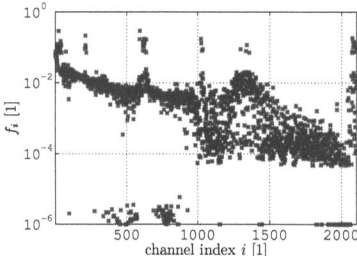
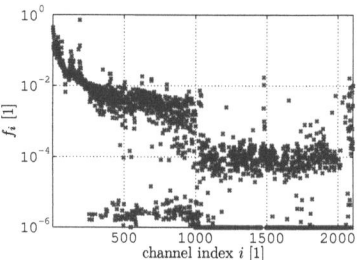

Figure 3.21.: Gain factors f_i of the decoupled control loops for the horizontal (left) and vertical (right) direction. All 2104 f_i are smaller than 1. The f_i have been artificially limited to a value of 10^{-6} to not create any open control loops.

way, since both directions are observed through the direction $\boldsymbol{v}[i]$. Since $\tilde{\boldsymbol{v}}[2i-1]$ and $\tilde{\boldsymbol{v}}[2i]$ are assumed to be independent, the effects of both directions can be added quadratically. By using further the fact that the luminosity depends quadratically on $\tilde{\boldsymbol{v}}[j]$, a combined luminosity effect for one controller direction can be calculated as $\Delta \mathcal{L}[i] = \tilde{G}_{\Delta\mathcal{L}}^2[2i-1] + \tilde{G}_{\Delta\mathcal{L}}^2[2i]$ and $\Delta\mathcal{L}_c[i] = \tilde{G}_{\Delta\mathcal{L},c}^2[2i-1] + \tilde{G}_{\Delta\mathcal{L},c}^2[2i]$. By considering that the luminosity loss only due to the offset $\Delta\mathcal{L}_o[i]$ can be approximated for small values by $\Delta\mathcal{L}_o[i] = \Delta\mathcal{L}[i] - \Delta\mathcal{L}_c[i]$, a modified cost function can be formulated as

$$\min_{f_i} \int_{\omega=-\infty}^{+\infty} \left[\frac{\Delta\mathcal{L}_c[i]}{\Delta\mathcal{L}[i]} \hat{Y}(\omega,f_i) + \frac{\Delta\mathcal{L}_o[i]}{\Delta\mathcal{L}[i]} \left| z^{-1} S_{IP,2}(e^{j\omega T_d}) \right|^2 \hat{Y}(\omega,f_i) \right] d\omega, \qquad (3.59)$$

where the z^{-1} models the behaviour of the actuator setpoint and beam-beam offset (IP system). The thereby calculated gain factors f_i are depicted in Fig. 3.21. Now that the f_i are defined the diagonal elements of $\boldsymbol{F}$ can be calculated as $f_i s[i]^{-1}$. Since the singular values $s[i]$ become very small for higher channels i, $f_i s[i]^{-1}$ tend to become large for high values of i. As a result the actuation of the tripods, due to control loops with high i is very large, even though the according luminosity loss is not very significant (see Fig. 3.20). Therefore, it turned out to be useful to limit the diagonal elements of $\boldsymbol{F}$, with $i > 300$, to a maximal value of 0.01. The according spatial filter has a Frobenius norm of 0.12, compared to a value of 1478.1 for the complete inversion of $\boldsymbol{R}$, which shows the strong BPM noise demagnification.

3.2.4. Discussion of the novel controller design method

After an extensive literature review, the option using as many actuators and sensors as possible for the orbit control was chosen. This choice ensures optimal beam oscillation damping, but makes the controller performance in general more sensitive to measurement noise and actuator imperfections. Given the actuators and sensors, two control structures were tested on the main linac. A decoupling design using the SVD algorithm was superior to a design using Kalman-predictors. The success of the SVD controller is based on the fact that the system to be controlled is very large and a simplification is necessary (with full system knowledge, the Kalman-predictor would the optimal solution). The

3.2. Linac controller

simplification by the decoupling of the SVD controller is especially effective, since in the decoupled control loops either the effect of ground motion or of BPM noise is dominant. Therefore, both signals can be treated differently by choosing different controller for different decoupled accelerator channels. Another reason for the effectiveness of the SVD controller is that the decoupling can be achieved for all frequencies, due to the simple dynamic structure of the accelerator system. SVD controller are hence especially well suited for the orbit control problem, which is consistent with the frequent use in storage rings.

This outcome gave rise to the development of the final, semi-automatic design procedure for the main linac and the BDS of CLIC. It consists of the following three steps.

1. Decoupling of inputs and outputs
2. Time-dependent filter design for the decoupled channels
3. Spatial filter design for the decoupled channels

In the first step the two matrices U and V are calculated with the SVD algorithm. By pre-multiplying the inputs with V and the outputs with U^T the system is decoupled. The large control problem can hence be split up into smaller sub-problems, which simplifies the design. In the second and third step, one individual controller is developed for each decoupled channel. To reduce the number of open parameters one equally parametrised time-dependent filter is used for each channel. Additionally one open gain parameter per channel (spatial filter) is used to minimise the channel output signal with respect to the excitation by ground motion and measurement noise. While the time-dependent filter is designed by the user manually, the determination of the open parameter is performed by an automatic minimisation algorithm. A second interpretation to the decoupling controller has been given. The decoupling matrices (U and V) and the gain parameters can be combined to form a matrix that can be interpreted as a measurement noise filter. It was found that the effectiveness of the noise filtering can be characterised by the reduction of the Frobenius norm of the measurement noise filter. In the case of CLIC the Frobenius norm was reduced from 1454.54 (no optimisation of the open parameters, but gains only set to one) to 0.19 (optimised parameters) in horizontal direction and from 1478.14 to 0.12 in vertical direction.

The presented algorithm has several advantages.

1. The generic design procedure can be adapted easily to other linacs and after a modification of the ground motion model also to storage rings.

2. The controller design makes it possible to incorporate models of the ground motion and the measurement noise. This closes the gap between the ground motion research of the accelerator community in the last decades and the orbit controller design practice.

3. The time-dependent filter enables the user to incorporate expert knowledge, e.g. element $P(z)$ in Sec. 3.2.3.2.

4. The tedious task of optimising each decoupled loop (2104 in the case of CLIC) by hand is taken over by an automated algorithm. This eases the task of the designer.

3. Controller design

5. The design time is significantly reduced.

6. As will be seen in Sec. 4.1, the controller performs better than a hand-optimised controller in the case of CLIC.

7. In Sec. 4.2, it will turn out that the controller is robust against imperfections, especially with respect to measurement noise.

8. Since the design is based on SVD decoupling, the overall controller is not too complicated and important insights are not lost.

3.3. Alternative designs for hardware cost reduction

In this section, two cost reduction options for the quadrupole stabilisation of CLIC are investigated, which are in the following named option 1 and option 2. The investigations are restricted to simplified, analytical estimates, which can be used to identify problems early and to get a first idea of necessary time constants of hardware components. Detailed full-scale simulations will have to verify the results presented. Additionally to the investigation of cost reduction options, a performance estimate for the baseline design, consisting of a simplified version of the L-FB (Sec. 3.2.3) in connection with the quadrupole stabilisation V1 or V2 (Sec. 1.3.2.4), will be given. The estimates for the baseline design have been an important information for the design of the L-FB.

The cost reduction option 1 aims to avoid the relatively expensive tripod system for the quadrupole stabilisation and the L-FB. Instead of moving the quadrupole mechanically, a dipole magnet is used as a corrector. The sensor measuring the quadrupole motion is now used in a feed forward fashion, since the quadrupole position itself is not changed by the quadrupole stabilisation anymore but only the magnetic centre of the quadrupole.

The cost reduction option 2 goes even one step further than option 1. It avoids not only the use of the tripod system, but also aims to reduce the number of corrector magnets. This does not result in a fundamental difference for the L-FB, but it does for the quadrupole stabilisation. Since not every quadrupole is equipped with a corrector anymore, the stabilisation task for these quadrupoles has to be taken over by the remaining correctors. Hence, the quadrupole stabilisation for the option 2 is a global algorithm similar to the L-FB, while the other quadrupole stabilisation options are correcting the motion of a quadrupole locally. For such a global algorithm network delays T_t are indispensible, since the sensor data have to be collected centrally and after some calculation have to be distributed to the correctors again. The delay time limits the bandwidth of the quadrupole stabilisation and consequently the performance at high frequencies.

The system structure for option 1 and option 2 is depicted in Fig. 3.22. To evaluate the effect of these options, but also of the simplified baseline design, the luminosity model presented in Sec. 2.2.3 is used. Using the according formulas the luminosity loss can be calculated as

$$\Delta \mathcal{L} = \frac{1}{(2\pi)^2} \iint_{-\infty}^{+\infty} \tilde{G}_{\Delta\mathcal{L}}^2(k) \left|S(e^{j\omega T_d})\right|^2 \left|S_{ST}(e^{j\omega T_{d,ST}})\right| P_{B10}(\omega,k) \mathrm{d}\omega \mathrm{d}k, \qquad (3.60)$$

3.3. Alternative designs for hardware cost reduction

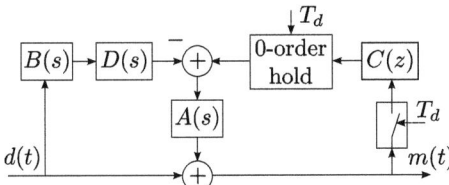

Figure 3.22.: The ground motion $d(t)$ moves the magnetic centre $m(t)$ of the quadrupole. This is counteracted in two ways. Ground motion measurements by a sensor with the transfer function $B(s)$ are used in the feed forward path $B(s)D(s)$, where $D(s)$ is the transfer function of the time delay due to the network communication and the limited processing speed, to reduce the change of the magnetic centre with the help of the actuator $A(s)$. The feedback path (L-FB), consists of a sampling element with sampling time T_d, a discrete controller $C(z)$ and a zero-order hold element, also modifies the magnetic centre via $A(s)$. In the depicted model the L-FB is assumed to act directly on the ground motion instead of the beam orbit, which is a simplification to the reality.

where $S_{ST}(e^{j\omega T_{d,ST}})$ and $S(e^{j\omega T_d})$ are the discrete frequency responses of the closed loop disturbance rejection transfer function of the quadrupole stabilisation and the sensitivity function of the L-FB. The term $\tilde{G}^2_{\Delta\mathcal{L}}(k)$ is the sensitivity function for luminosity (stabilised version, see Sec. 3.2.3). Note that the used luminosity model neglects the spatial properties of the L-FB. Hence, for option 2 only the effects due to the introduced delay time T_t can be studied, but not the performance reduction due to reduction of the number of used correctors (worse beam steering).

In the case the baseline option is used, $S_{ST}(e^{j\omega T_{d,ST}})$ is given numerically by the curves in Fig. 1.8 (right). If, on the other hand, the feed forward options are investigated, the frequency response is given by

$$S_{ST}(j\omega) = 1 - B(j\omega)D(j\omega)A(j\omega) \quad \text{with} \quad (3.61)$$
$$D(j\omega) = e^{-j\omega T_t}, \quad (3.62)$$

where $A(j\omega)$ and $B(j\omega)$ are the frequency responses of the used actuator and sensor. The expression $D(j\omega)$ in Eq. (3.62) corresponds to a time delay by T_t. Two types of sensors are tested: the CMG-6T seismometer from Guralp Systems $B_1(s)$ and a proposed geophone called L4C $B_2(s)$. Note that the baseline quadrupole stabilisations version 1 and 2, use the same sensors as the feed forward options, but in a feedback configuration. The frequency responses of $B_1(s)$ and $B_2(s)$ are given in numerical form and are plotted in Fig. 3.23. Also two different types of actuators,

$$A_1(s) = \frac{1}{1+T_1 s} \quad \text{and} \quad A_2(s) = \frac{1}{(1+T_1 s)^2}, \quad (3.63)$$

were tested. The first actuator $A_1(s)$ corresponds to a first order low pass, while $A_2(s)$ is a second order low pass. Since $A_2(s)$ has two equal real poles, it models the case of adiabatic damping. The parameter T_1 is the time a first order low pass needs to reaches 63.2 % of the amplitude of an applied step function.

3. Controller design

Contrary to the feed forward transfer function, the disturbance rejection function of the L-FB is a discrete-time function given by

$$S(z) = \frac{1}{1 + C(z)A(z)}, \quad \text{with} \tag{3.64}$$

$$C(z) = \frac{z}{z-1}, \tag{3.65}$$

where $C(z)$ is the controller transfer function and $A(z)$ the discretised form of the continuous actuator $A(s)$. As can be seen from Eq. (3.65), only a simple integrator is used as a controller. This is a simplification compared to the time-dependent filter $g(z)$ designed in Sec. 3.2.3.2. The reasoning behind this simplification is that the actuator $A(z)$ takes over the role of the low pass $L(z)$. The elements $P(z)$ and $E(s)$ included in $g(z)$ are also neglected, since their impact on the transfer function is small. No BPM noise is included in the estimations.

The sampled $A(z)$ can be calculated from $A(s)$ with the help of the Z-transform. This transform was already introduced for the design of the low pass $L(z)$ in Sec. 3.2.3.2. In the same section, the Z-transform of $A_1(s)$ was already derived and is restated here as

$$\frac{\left(1 - e^{-\frac{T_d}{T_1}}\right)}{z - e^{-\frac{T_d}{T_1}}} \quad \text{with} \quad T_d = 0.02\,\text{s}. \tag{3.66}$$

The Z-transform of $A_2(s)$ is calculated by the formula

$$A_2(z) = (1 - z^{-1}) Z\left\{\frac{1}{s(1+sT_1)^2}\right\}. \tag{3.67}$$

To be able to evaluate Eq. (3.67) the argument of $Z\{.\}$ has to be expanded into partial fractions

$$A_2(s) = \frac{1}{(1-T_1s)^2} = \frac{1}{s} + \frac{1}{s + \frac{1}{T_1}} - \frac{1}{T_1}\frac{1}{(s+\frac{1}{T_1})^2}. \tag{3.68}$$

Using Eq. (3.68) in Eq. (3.67) and considering that the Z-transform is a linear operation, $A_2(z)$ can be evaluated with a table look-up and after short calculation to

$$A_2(z) = \frac{1 - e^{-\frac{T_d}{T_1}}}{z - e^{-\frac{T_d}{T_1}}} - \frac{T_d}{T_1}\frac{(z-1)e^{-\frac{T_d}{T_1}}}{(z - e^{-\frac{T_d}{T_1}})^2}. \tag{3.69}$$

In Figs. 3.24 and 3.25, the results of the evaluation of Eq. (3.60) for different parameters are presented. Figure 3.24 shows the estimated luminosity loss due to different actuator dynamics for the baseline design with quadrupole stabilisation V1 or V2. Since the frequency responses of the baseline quadrupole stabilisation versions are only given in closed form, the effect of the actuator dynamics is only taken into account for the L-FB. Obviously, the quadrupole stabilisation V2 produces a much smaller luminosity loss than V1. This result is consistant with the full-scale simulations presented in Sec. 4.1. The quadrupole stabilisation V1 produces the best results with a first-order low pass

3.3. Alternative designs for hardware cost reduction

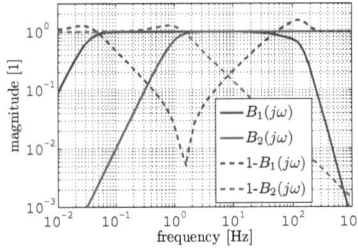

Figure 3.23.: The magnitude of the sensor frequency responses $B_1(j\omega)$ and $B_2(j\omega)$ can be used for feed forward control. If no network delay is considered ($T_t = 0$) and the actuator is very fast compared to the time constants of the sensor ($T_1 = 0$), the feed forward frequency response can be calculated as $1 - B(j\omega)$.

Figure 3.24.: Estimated luminosity loss due to ground motion for the simplified baseline design for either the quadrupole stabilisation V1 or V2. Different actuator types $A_1(s)$ and $A_2(s)$ and actuator time constants T_1 are investigated.

actuator with $T_1 = 0.1\,\text{s}$, even though BPM noise is not taken into account. In the full design of the L-FB, this result is considered by using the low pass $L(z)$ as an element of the time-dependent filter $g(z)$ in Sec. 3.2.3.2. The result also shows that the actuator dynamics are not as crucial for the L-FB as initially expected and that a slower actuator is even beneficial for the luminosity performance. It is further interesting to see, that the use of a second order low pass $A_2(s)$ can lead to performance problems for $T_1 \approx 40\,\text{ms}$. The problem arises due to the fact that the actuator has a too large phase advance and causes thereby instabilities. The phase of the actuator is therefore more important than the time constant T_1, which should be considered for the design the actuators.

In Fig. 3.25 (left), the cost reduction option 1 is investigated. Since the actuator is now also taken into account for the quadrupole stabilisation, the luminosity performance

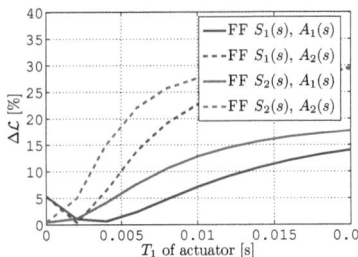
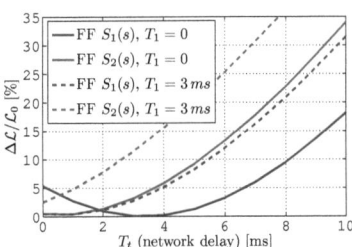

Figure 3.25.: Luminosity loss estimate due to ground motion for the cost reduction options 1 (left) and 2 (right) and a simplified version of the L-FB.

3. Controller design

is much more sensitive to T_1. The Guralp seismometer $B_1(s)$ achieves better results if compared to the proposed geophone $B_2(s)$. Note that this is in contrast to the results for the baseline versions in Fig. 3.24, in which the feedback system based on the proposed geophone performed better than the feedback system using the Guralp seismometer. The luminosity peaks due to the instabilities in Fig. 3.24 are not visible in Fig. 3.25, since they occur not before T_1 of about 40 ms, which exceeds the scale of the plot. It can be concluded, that for option 1, the sensor $B_1(s)$ works better than $B_2(s)$ and that the actuator $A_1(s)$ is preferable to $A_2(s)$. A time constant T_1 of 0 to 10 ms would be acceptable.

Finally, the cost reduction option 2 is investigated in Fig. 3.25 (right). Only the actuator $A_1(s)$ is considered with two different time constants $T_1 = 0$ ms and $T_1 = 3$ ms. It is not surprising that the actor with $T_1 = 0$ ms produces better results than the one with $T_1 = 3$ ms. Also the sensor $B_1(s)$ performs better than $B_2(s)$, which has already been observed for option 1. For the best performing configuration, using $B_1(s)$ and $A_1(s)$, a network delay T_t of 0 to 8 ms is acceptable.

Concluding, one can say that the estimates presented in this section indicate that both cost reduction options are capable of preserving the luminosity to an acceptable level. This results have to be checked with more detailed simulation to verify the validity of the made assumptions.

3.4. IP controller

In this section, we will describe control algorithms for the IP feedback (IP-FB). The following explanations assume that the reader is familiar with the functional principle of the IP-FB as presented in Sec. 1.3.2.2. The final version of the IP-FB is developed by a collaboration of the institutes LAPP and SYMME (see Caron et al. [20]). The feedback controller presented in this section is a complementary design. While the LAPP-SYMME controller is a complex algorithm, the presented feedback strategies are simpler but still effective. These simpler designs allow to change the feedback algorithms fast and therefore to respond quickly to system changes. This is especially important for the optimisation phase of the different mitigation methods with respect to each other, where parameter changes occur often.

An introduction to the purpose and structure of the IP feedback was already given in Sec. 1.3.2.2. To rephrase shortly, the task of the IP feedback is to reduce the beam-beam offset at the IP. This offset is induced mainly by a misalignment of the final doublets (FD) of the electron and positron BDS. The main cause of this misalignment is ground motion. Since the beam-beam offset is much more sensitive to the misalignment of magnets in the FD than to misalignment of other magnets, a special counter measure has been designed: a mechanical pre-isolator (see Sec. 1.3.2.5). This pre-isolator is a huge mass-spring system, which damps high-frequent ground motion efficiently.

For the design of the controller in this section, it is essential to have a model of the spectrum of the beam-beam offset. Such a model has been developed in Sec. 2.3.1 and includes the ground motion spectrum, the frequency response of the pre-isolator and the transfer of the beams from the magnets of the FD to the IP. Including in this model also the effect of measurement noise with a PSD $N_{IP}(\omega)$, the beam jitter produced before

3.4. IP controller

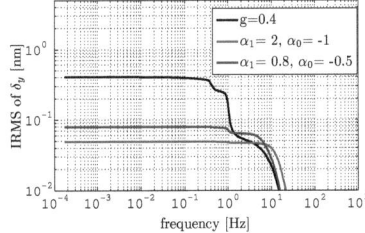
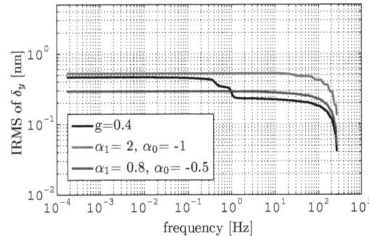

Figure 3.26.: Comparison of the integrated RMS of the beam-beam offset δ_y for three IP-FBs. On the left side no upstream jitter and no BPM noise is used. Both double integrator controller (blue and red) outperform the integrator controller (black) easily. In the right side upstream beam jitter ($\sigma_d = 0.2$ nm) and BPM jitter ($\sigma_n = 20$ pm) are included in the simulations. The performance of the double integrator feedback degraded. Notice that the best controller without disturbances and noise (red curve left) is the worst with them (red curve right).

the FD with a PSD $D_{IP}(\omega)$, the variance of the beam-beam offset $\sigma_\delta^2 = E\{y_\delta^2\}$ at the IP can be written as

$$\sigma_\delta^2 = \int_{-\infty}^{+\infty} |T_{IP}(e^{j\omega T_d})|^2 N_{IP}(\omega) + \int_{-\infty}^{+\infty} P(\omega, k) |G_\delta(j\omega, k)|^2 \left|S_{IP}(e^{j\omega T_d})\right|^2 \frac{dk}{2\pi}$$
$$+ D_{IP}(\omega) \left|H_{IP}(e^{j\omega T_d}) S_{IP}(e^{j\omega T_d})\right|^2 \frac{d\omega}{2\pi}. \tag{3.70}$$

The term $G_\delta(\omega, k)$ is the sensitivity function for the beam-beam offset as calculated in Sec. 2.3.1. The terms $S_{IP}(e^{j\omega T_d})$ and $-T_{IP}(e^{j\omega T_d})$ are the sensitivity function and noise frequency responses of the IP-FB and $H_{IP}(e^{j\omega T_d}) = e^{-j\omega T_d}$ is the frequency response of system to be controlled. We will in the following only consider the vertical beam-beam offset. The horizontal direction can be treated analogously.

The upstream beam jitter and the measurement noise are modelled as white, Gaussian stochastic processes with zero mean. Thus $D_{IP}(\omega, k)$ and $N_{IP}(\omega, k)$ are constant PSDs, which amplitudes are determined by the assumed standard deviations σ_d and σ_n. From simulations an estimate for σ_d of 20 pm was obtained, where the result depends strongly on the ground motion model, the L-FB and the stabilisation system used. Due to the beam-beam offset measurement setup (see Sec. 1.3.2.2), σ_n is as small as 20 pm. This extremely small value of σ_n is one of the major differences between the IP-FB and the L-FB. While the main focus of the L-FB was the reduction of BPM noise effects, this is of less importance for the IP-FB. The noise level is so small that the IP-FB can be even more optimised for ground motion suppression. Also the fact that the IP-FB is a single-input, single-output system eases the design.

In the following two simple IP-FB algorithms are presented. Both do not use the measurements δ directly, but apply as a first step the inverse of the beam-beam deflection curve (see Fig. 1.6). This step linearises the problem. The corrected measurements $\hat{\delta}$

3. Controller design

are then processed by the two discrete-time control algorithms

$$u_1(z) = \frac{gz}{z-1}\hat{\delta}(z) = C_1(z)\hat{\delta}(z) \quad \text{and} \tag{3.71}$$

$$u_2(z) = \frac{z(\alpha_1 z + \alpha_0)}{(z-1)^2}\hat{\delta}(z) = C_2(z)\hat{\delta}(z). \tag{3.72}$$

The controller $C_1(z)$ is a simple integrator with a variable gain g. It is similar to the time-dependent algorithm in Sec. 3.2.2.2 and has the same physical motivation. The second algorithm $C_2(z)$ uses a double integrator with two open parameter that correspond to a constant gain and the position of the zero of the controller. Considering that the plant transfer function is $H_{IP}(z) = 1/z$ and using the standard expressions for transfer functions in Appendix C.2, the disturbance and noise transfer functions have the form

$$S_{IP,1}(z) = \frac{z-1}{z+(g-1)}, \qquad S_{IP,2}(z) = \frac{(z-1)^2}{z^2 + z(\alpha_0 - 2) + (\alpha_0 + 1)} \tag{3.73}$$

$$-T_{IP,1}(z) = -\frac{g}{z+(g-1)} \quad \text{and} \quad -T_{IP,2}(z) = -\frac{\alpha_1 z + \alpha_0}{z^2 + z(\alpha_0 - 2) + (\alpha_0 + 1)}. \tag{3.74}$$

Both, $S_{IP,1}(z)$ and $-T_{IP,1}(z)$ are stable for $0 < g < 2$. For the stability analysis of $S_{IP,2}(z)$ and $-T_{IP,2}(z)$ we use the conditions in Eqs. (3.18) and (3.19). This leads to the three stability conditions $-2 < \alpha_0 < 0$, $\alpha_1 > -\alpha_0$ and $\alpha_1 < \alpha_0 + 4$. The parameter space for all stable controller for $C_2(z)$ is hence of triangular shape in the α_0-α_1 plane. This allowed parameter space was scanned to find the controller that minimises the beam-beam offset prediction in Eq. (3.70).

In Fig. 3.26 the resulting beam-beam offsets for different controller configurations are depicted. The final performance changes strongly depending on the standard deviation of the upstream beam jitter. Since this jitter is assumed to be white, every controller will always amplify this value. The jitter standard deviation is therefore a lower bound for the possible IP-FB performance. Considering this fact, the performance of $\delta_y = 0.3\,\text{nm}$, of the controller corresponding to the black curve in Fig. 3.26 (right), is close to the theoretical optimum. To further improve the result a Kalman-predictor as in Sec. 3.2.2.1 could be used. Since the measurement noise level is much smaller at the IP, the use of such a filter is more promising than for the linac feedback. Such a filter would also allow to include a more realistic model of the upstream beam jitter.

4. Controller performance, imperfections and robustness studies

In this chapter, the performance of the L-FB (designed in Chap. 3) in combination with the other three ground motion mitigation methods (quadrupole stabilisation, IP-FB and pre-isolator) is evaluated. For this evaluation, the simulation framework presented in Sec. 2.4 is used. The effect of different imperfections, as for example ground motion, BPM noise and acceleration gradient jitter, is calculated.

As a performance and robustness measure the averaged peak luminosity is used. To prevent misunderstandings in the interpretation of the presented plots, it should be mentioned that CLIC is over-dimensioned by design to produce a luminosity of about 120 %. The additional 20 % account for the luminosity loss due to dynamic imperfections as e.g. ground motion.

4.1. Performance in presence of ground motion effects

In this section the effect of ground motion on the accelerator performance is investigated. In Fig. 4.1, the relative luminosity is shown when ground motion of model B10 is applied and different combinations of mitigation methods are used. In case no mitigation methods are used at all (red curve left) hardly any luminosity is produced. When turning on the IP-FB (IP controller according to Eq. (3.72)), slow beam-beam offset is reduced and the luminosity performance is improved (blue curve left). The high-frequency offset can be reduced by adding the pre-isolator (green curve left) and the quadrupole stabilisation (blue curve right), where the quadrupole stabilisation version 1 is used in this case. Even though the quadrupole stabilisation reduces the beam-beam offset, it also causes a significant low-frequency increase of the beam size at the IP, which leads to strong luminosity loss. This beam size growth can be reduced by using the L-FB additionally to the other mitigation methods (red curve right).

By systematic modification of the frequency response of the quadrupole stabilisation version 1 in the simulations, it was found that the remaining luminosity loss could be reduced, if the two peaks of the frequency response (see Fig. 1.8 (right)) would be shifted to higher frequencies. This improvement is due to several reasons. The lower peak of the frequency response of version 1 at about 0.3 Hz causes an especially strong amplification of ground motion, since it amplifies parts of the microseismic peak. If it is shifted to a frequency of about 1 Hz, the ground motion excitation is reduced and at the same time the mismatch between the transfer functions of the quadrupole stabilisation and the pre-isolator is lowered. The second peak of the quadrupole stabilisation function version 1 is located at about 75 Hz, where the frequency components of the ground motion are still large enough to cause significant luminosity loss. By shifting this peak to e.g. 100 Hz, the excitation is reduced. Additionally, the L-FB is very efficient for frequencies that

4. Controller performance, imperfections and robustness studies

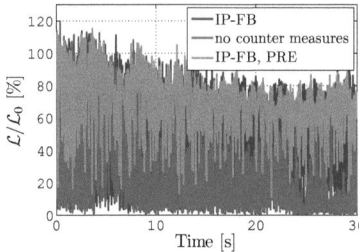 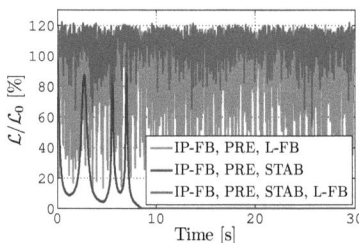

Figure 4.1.: Relative luminosity $\mathcal{L}/\mathcal{L}_0$ due to ground motion of model B10 with different ground motion mitigation methods. In the depicted curves, no averaging over different seeds of the random generator has been performed, in order to give an impression of the real-time accelerator behaviour. A discussion of the curves is given in the text.

	A	B	B10	C
V1	8.7 %/4.0 %	9.3 %/5.1 %	15.8 %/11.7 %	120.3 %/100.8 %
V2	1.4 %/0.0 %	1.6 %/0.5 %	2.0 %/0.8 %	67.6 %/71.5 %

Table 4.1.: Relative luminosity loss $\Delta\mathcal{L}/\mathcal{L}_0$ for different ground motion model (columns) and stabilisation systems (rows). For every combination, two relative luminosity loss values are listed. The first corresponds to a hand-optimised spatial L-FB, which is the same for every combination. The second listed value originates from simulations with the automatically optimised spatial L-FB, which is individual for every combination. Each listed value corresponds to an average over 20 ground motion seeds simulated for a machine operation of 30 s.

are a multiple of 50 Hz, and therefore the amplification of the quadrupole stabilisation system at 100 Hz is compensated by the L-FB.

These observations were used by the quadrupole stabilisation team to create a new quadrupole stabilisation system that is referred to as version 2 (see Janssens et al. [59]). Contrary to version 1, version 2 uses a proposed geophone as a sensor, which is currently under development. This geophone is better suited for higher frequencies compared to the seismometer CMG-6T used for version 1. The predicted frequency response of the quadrupole stabilisation version 2 is shown in Fig. 1.8 (right).

To evaluate the combined effect of all mitigation methods, the relative luminosity loss due to ground motion of different models has been simulated. For the L-FB, hand-optimised controller gains f_i as well as automatically optimised gains (see Sec. 3.2.3.3) have been tested. For the quadrupole stabilisation, version 1 and 2 have been considered. A BPM resolution of 100 nm in the main linac and 50 nm in the BDS has been presumed. The results are summarised in Tab. 4.1. The figures show that the automatically optimised L-FB is superior compared to the hand-optimised version. Also the quadrupole stabilisation version 2 performs better than version 1. Apart from the ground motion model C, which is a very pessimistic assumption for the future CLIC site, the luminosity requirements of CLIC can be met for all cases.

4.2. Imperfections and robustness

In the following, the effects of different imperfections on the performance of the L-FB and the IP-FB are investigated (see also Pfingstner et al. [88]). To be able to compare the impacts of the different imperfection, the same mitigation method configuration is used for all simulations. If not stated differently, the L-FB uses a spatial filter with gains optimised for ground motion model B10 and the quadrupole stabilisation version 1. Additionally, the quadrupole stabilisation version 1 and the double integrator IP-FB (see Eq. (3.72)) are applied. In the following plots each data point corresponds to an averaged luminosity over 200 time steps (4 s).

4.2.1. BPM resolution

To analyse the effect of the limited BPM resolution, white, Gaussian noise of a certain variance was added to the BPM measurements. The relative luminosity loss resulting from a limited resolution of the BPMs used by the L-FB is depicted in Fig. 4.2. Different versions of the L-FB are investigated and ground motion is not applied.

When using a simple dead-beat controller (integrator for the time-dependent filter $g(z)$ and the identity matrix for the spatial filter $\boldsymbol{F}$), the luminosity loss is very sensitive to the BPM resolution (black curve). Note that even for very small BPM resolutions the luminosity loss is about 8 %. This is due to the fact that the L-FB picks up unavoidable simulation noise that is caused by the limited number of simulated particles in the simulated beam (Schottky noise).

The noise behaviour of the L-FB can be significantly improved by optimising the gain factors f_i, which are collected in $\boldsymbol{F}$, with respect to the ground motion excitation (red curve). A further improvement can be achieved, if additionally to the integrator also the low pass $L(z)$, the peak $P(z)$ and the phase lifting element $E(z)$ are used (see Sec. 3.2.3.2) for the time-dependent filter (blue curve). This configuration of the L-FB results in an acceptable luminosity loss of 2 % for a BPM resolution of 50 nm. The main contribution to this luminosity loss originates from the BPM resolution in the BDS. This can be inferred from the green curve, where a perfect BPM resolution in the BDS was used. Due to these observations the specifications for the BPM resolution could be relaxed, from initially 10 nm for all BPMs, to 50 nm for the BPMs in the BDS and 100 nm for the BPMs in the main linac. These larger tolerances lead to a significant cost reduction.

In Fig. 4.3, the luminosity loss as a function of the resolution of the post-collision line BPM used for the IP-FB is shown for different IP-FB versions. Two controller transfer functions are investigated: the simple integrator feedback $C_1(z)$ with variable gain factor g (see Eq. (3.71)), and the double integrator feedback $C_2(z)$ with the parameters $\alpha_0 = -0.5$ and $\alpha_1 = 0.8$ (see Eq. (3.72)). The simulations show that for all of these IP-FBs the luminosity loss due to a limited BPM resolution is negligible up to a BPM resolution of about 10 μm. Since the post-collision line BPM is assumed to have a resolution of about 1 to 3 μm, this resolution is not a critical design issue.

101

4. Controller performance, imperfections and robustness studies

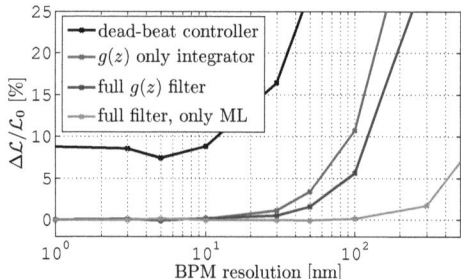

Figure 4.2.: Relative luminosity loss $\Delta\mathcal{L}/\mathcal{L}_0$ due to the resolution of the BPMs in the main linac and BDS. The black curve corresponds to the use of a dead-beat controller for the L-FB, which means that for the time-dependent controller $g(z)$ only a integrator is used and $\boldsymbol{F} = \boldsymbol{I}$ (see Sec. 3.2.3 for details). For the red curve a controller also using a simple integrator for $g(z)$ but automatically optimised gains f_i for the spatial filter (ground motion model B10 and quadrupole stabilisation system V1) is switched on. For the blue curve, the time-dependent filter $g(z) = I(s)L(z)P(z)E(z)$ and the automatically optimised gains f_i have been used, which corresponds to the baseline L-FB. Also the green curve has been obtained with the baseline L-FB, but perfect BPMs in the BDS have been assumed.

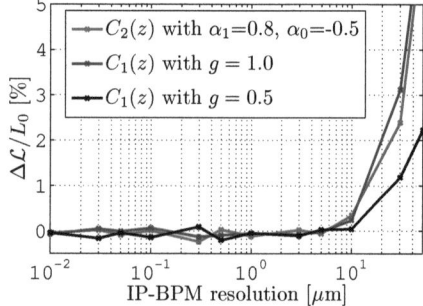

Figure 4.3.: Relative luminosity loss for different IP-FBs variants due to the resolution of the BPM in the post-collision line. The labels $C_2(z)$ and $C_1(z)$ correspond to different IP controller as defined in Sec. 3.4. For the $C_1(z)$ two different controller gains g are tested. Up to a resolution of $10\,\mu m$ hardly any luminosity loss is observed for all types of IP-FBs. The small fluctuations below a resolution of $10\,\mu m$ come from the numerical simulation noise of the used beam-beam code GUINEA-PIG.

4.2.2. Energy imperfections, dispersion filter and energy measurement

Deviations of the beam energy from the design value can be caused by unwanted changes of the acceleration voltages in the structures of the main linac, but also by initial beam energy jitter at the entrance of the main linac. The main cause for changes of the acceleration gradients are deviations of the amplitude and the phase of the drive beam from its specified values. As can be seen in Fig. 4.4 (left), variations of acceleration gradients can lead, in combination with the L-FB operation, to large luminosity loss (blue curve). The reason for this loss is that the beam energy deviations cause large beam offsets in the horizontal direction due to the dispersion in the BDS (more precisely in the so called collimation section). These large beam offsets are measured by the BPMs and coupled back via the L-FB. As a result the beam is steered incorrectly, which leads to strong luminosity loss.

To counteract this problem, a simple strategy named *dispersion filter* was developed. The dispersion filter removes the dispersive orbit from the BPM measurements. Only the filtered BPM measurements $\tilde{x}$ are used as an input for the L-FB. To remove the offsets caused by the energy deviations, the dispersion filter uses the shape of the dispersive orbit in the BPMs, which was determined via simulations and stored in the vector x_D. The offsets due to energy variations can be removed from the BPM measurements x by the simple procedure

$$\tilde{x}[k] = x[k] - f_D[k] x_D \quad \text{with} \tag{4.1}$$

$$f_D[k] = \frac{x[k]^T x_D}{x_D^T x_D}, \tag{4.2}$$

where k is the time step index. The use of the dispersion filter leads to a significant improvement of the L-FB performance, which is depicted in the red curves in Fig. 4.4 (left) and (right). The additional luminosity loss due to the L-FB (blue curve in Fig. 4.4 (right)) is seen to be below 0.5 % up to a gradient jitter with a standard deviation of 0.5 %, and therefore small compared to the loss without L-FB. If the dispersion filter is used in connection with ground motion and no beam energy imperfections are applied, the luminosity decrease due to the filtering of the dispersive orbit is only in the order of 0.1 % and hence negligible.

Similar to the beam energy deviation due to a change of the acceleration gradients, also an initial beam energy jitter causes large dispersive beam offsets that couple to the L-FB (see Fig. 4.5 (left)). By using the dispersion filter this problem can be overcome and the luminosity loss up to an initial energy jitter of 4 % is seen to be negligible.

The dispersion filter can not only be used to filter dispersive orbits from the BPM measurements, but potentially also for precise energy measurements. Figure 4.5 (right) shows that the relation between beam energy deviations at the end of the main linac and the factor f_D is linear over a large range. Simulations were performed in which the beam energy deviation was calculated from f_D under the assumption that the factor between f_D and the beam energy as well as the shape of the dispersive orbit is known. Errors of the acceleration gradient amplitudes and phases were introduced, which resulted in a beam energy jitter at the end of the main linac of 0.13 % (standard deviation). With the help of the dispersion filter, the beam energy at the end of the main linac could be reconstructed from the BPM measurements with a relative accuracy of about 2×10^{-5}. This

4. Controller performance, imperfections and robustness studies

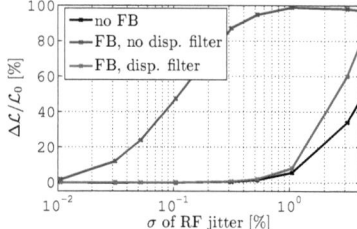 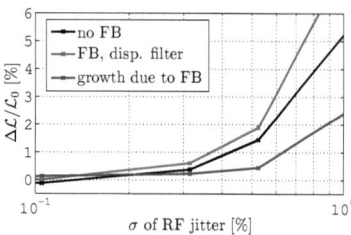

Figure 4.4.: Relative luminosity loss $\Delta\mathcal{L}/\mathcal{L}_0$ due to white, Gaussian jitter of the acceleration gradients of each decelerator with a relative standard deviation of σ. (left) If the L-FB is used without modifications, the luminosity is strongly decreased due to the large dispersive beam offset caused by the beam energy jitter (blue curve). If additionally to the L-FB the dispersion filter is used, the luminosity loss is just slightly above the loss without L-FB. (right) The additional luminosity loss due to the L-FB is below 0.5 % up to an acceleration gradient jitter of 0.5 %.

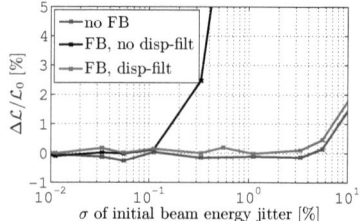 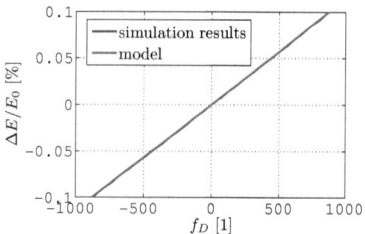

Figure 4.5.: (left) Relative luminosity loss $\Delta\mathcal{L}/\mathcal{L}_0$ due to white, Gaussian beam energy jitter at the entrance of the main linac with a relative standard deviation σ. (right) The relation between the factor f_D of the dispersion filter and the relative energy deviation $\Delta E/E_0$ at the end of the main linac is linear over a wide range of values. A model (red curve) of the form $\Delta E/E_0 = 1.138 \times 10^{-4} f_D$, where $\Delta E/E_0$ is given in percent, fits the simulated results (blue curve) very accurately.

4.2. Imperfections and robustness

value is limited by the accuracy with which the shape of the dispersive orbit is known, but not by the BPM resolution and ground motion disturbances. The insensitivity with respect to the BPM resolution can be explained by considering that a beam energy deviation of 0.1 % results in a value for f_D of about 1000, while the projection of a BPM noise vector results only in a value of about 0.05. Hence, the dispersion filter provides a potentially very precise energy measurement, which accuracy is limited in practical applications mainly by the accuracy with which the shape of the dispersive orbit and the constant factor connecting f_D and the beam energy deviation can be measured.

4.2.3. Quadrupole position errors

In this section the effect of position errors of quadrupoles is evaluated. Such position errors can originate from two main sources. The first error source is the noise of the sensor of the quadrupole stabilisation system that is fed back by the feedback controller and and acts in that way on the quadrupole position. The second source originates from imperfection of the quadrupole stabilisation system with respect to its positioning capabilities. When the L-FB changes the set point of the quadrupole stabilisation system, limited DAC resolution or difficult to control actuator dynamics can lead to an error between the real and the demanded quadrupole position. In the following, effects of both, the quadrupole stabilisation sensor noise and the limited positioning accuracy, will be analysed.

At first, the effect of the limited positioning accuracy is investigated. White noise was used to alter the quadrupole positions and the resulting luminosity loss was recorded with and without the action of the L-FB. Several conclusions can be drawn from the simulation results in Fig. 4.6. It can be seen that the L-FB worsens the luminosity loss by approximately a factor two. This result was expected, since every feedback controller amplifies the effect of a white disturbance (no prediction possible). The positioning errors of quadrupoles in the BDS are observed to be more significant for the luminosity loss than the errors for quadrupoles in the main linac. This observation suggests that it could be favourable to have two quadrupole stabilisation systems in order to reduce the costs; one with more demanding specifications for the BDS, and a cheaper one for the main linac. Another important observation is that the luminosity loss increases strongly, if the quadrupole QF1 of the final doublet is also misaligned. This quadrupole (and also QD0) is very sensitive to misalignments and hence the decision was taken to not use it as a corrector for the L-FB. Finally, it can be stated that for a luminosity loss smaller than 0.5 %, a Gaussian positioning error of the quadrupole stabilisation system should not exceed a standard deviation of 0.25 nm. This tolerance is tight, but is within reach considering that the actuation steps of the L-FB are in the same range.

The effect of uniformly distributed quadrupole positioning errors is investigated in Fig. 4.7. Such an error distribution is typical, if the error is dominated by the limited resolution of the digital to analogue converter (DAC) of the controller electronics. The positioning errors take in this case values between $-\Delta/2$ and $+\Delta/2$. The main outcome of the results in Fig. 4.7 is that the luminosity loss due to uniformly distributed quadrupole position errors can be very well approximated by the luminosity loss due to Gaussian distributed errors. The fact has to be used that a uniform distribution can be approximated by a Gaussian distribution with a standard deviation of $\sigma = \Delta/\sqrt{12}$.

4. Controller performance, imperfections and robustness studies

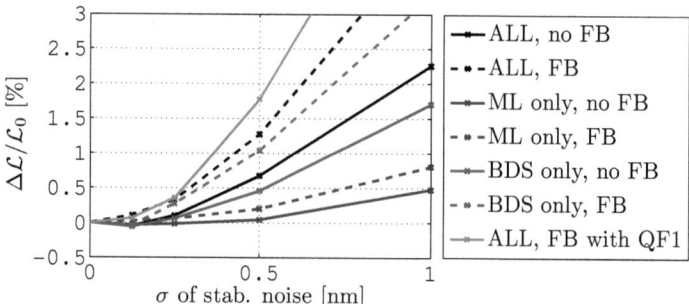

Figure 4.6.: Relative luminosity loss $\Delta\mathcal{L}/\mathcal{L}_0$ due to white, Gaussian jitter of the corrector actuations in the main linac, the BDS and both, with a standard deviation σ. The dashed curves correspond to simulations in which the L-FB and the IP-FB have been turned on, while for the solid curves these feedback systems have been turned off. For comparison, the green line shows the effect, if the quadrupole QF1 (final doublet) is also misaligned.

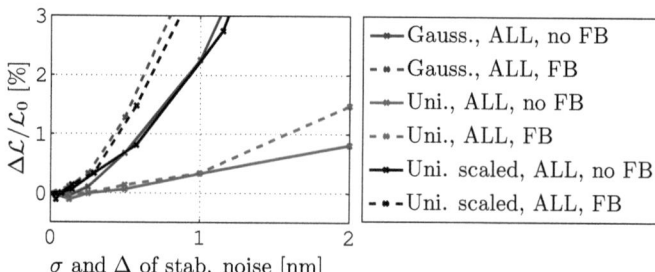

Figure 4.7.: Relative luminosity loss $\Delta\mathcal{L}/\mathcal{L}_0$ due to uniformly distributed quadrupole position errors (red curves) taken from the interval $[-\Delta/2, +\Delta/2]$ for main linac and BDS. If the results of the uniformly distributed errors are scaled by a factor $1/\sqrt{12}$ (black curves) they are very similar to the results of the Gaussian distributed errors (blue curves). This originates from the fact that a uniform distribution can be approximated by a Gaussian distribution with a standard deviation of $\sigma = \Delta/\sqrt{12}$

4.2. Imperfections and robustness

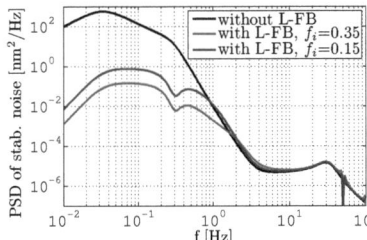
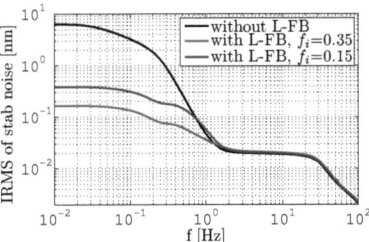

Figure 4.8.: PSD (left) and IRMS (right) of the quadrupole motion due to the noise induced by the sensor of the quadrupole stabilisation system version 1. The black curve corresponds only to the noise from the quadrupole stabilisation, while for the red and the blue curve the effect of of the L-FB has been considered for two different effective gain factors f_i.

After evaluating the effect of the limited positioning accuracy of the quadrupole stabilisation system, we focus now on the error induced by the sensor of the quadrupole stabilisation system. The quadrupole stabilisation group provided us an estimate of the PSD of the quadrupole motion due to the sensor noise for the quadrupole stabilisation system version 1 (see black curve in Fig. 4.8 (left)). The IRMS of this PSD (see black curve in Fig. 4.8 (right)) indicates a motion of about 7 nm. This motion would cause a large luminosity loss, if it would be a white stochastic process. It can be seen however that the quadrupole motion spectrum is by no means white.

To get a more realistic estimate of the luminosity loss, the original PSD is multiplied with the squared sensitivity function of the L-FB $|\hat{S}(e^{j\omega T_d})|^2$ (defined in Eq. (3.52)) to include the action of the orbit feedback into the estimation. Since the L-FB uses different gain factors f_i for different spatial directions, it is not obvious which effective gain factor should be used for the calculation of $\hat{S}(e^{j\omega T_d})$. Thus, two different f_i are investigated, where the choice of $f_i = 0.35$ assumes that most of the luminosity loss is caused by the first few directions of the L-FB and $f_i = 0.15$ represents the case where approximately the first 100 directions have a significant impact. When the original PSD is folded with the according L-FB sensitivity functions, the resulting IRMS motion is between 0.2 and 0.4 nm. This corresponds to a luminosity loss of about 0.1 to 0.5 %. Note that for this evaluation the solid black curve in Fig. 4.6 has to be used, since the action of the L-FB has already been taken into account. The luminosity loss due to the noise of the quadrupole stabilisation system is therefore in an acceptable range.

4.2.4. Other imperfections

In this section, the luminosity loss due to errors of the magnetic strengths of the quadrupoles, scaling error of BPMs and correctors, initial beam jitter and BPM noise in the orbit response matrix R used for the L-FB is considered. These imperfections have been evaluated with and without the action of the L-FB.

Figure 4.9 shows the luminosity loss due to white, Gaussian errors of the quadrupole

4. Controller performance, imperfections and robustness studies

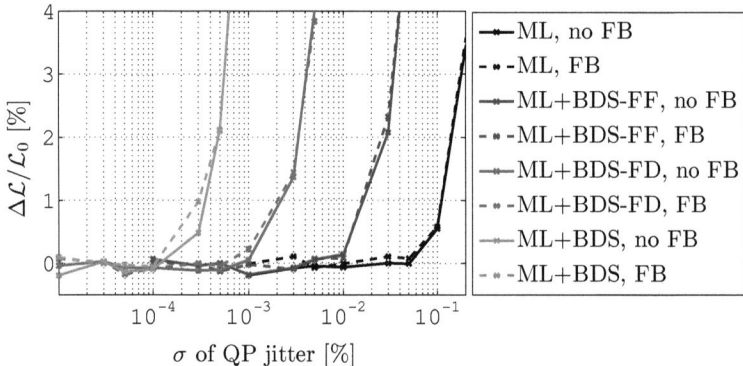

Figure 4.9.: Relative Luminosity loss $\Delta\mathcal{L}/\mathcal{L}_0$ due to white, Gaussian jitter of the magnetic strengths of the quadrupoles, with a standard deviation of σ. Different parts of CLIC have been considered, where FF stands for final focus and FD for final doublet. The dashed curves correspond to simulations in which the L-FB and the IP-FB have been turned on, while for the solid curves these feedback systems have been turned off.

magnet strengths. No static misalignments and ground motion have been used for this study. Orbit changes occure only in the BDS, where the varying quadrupole magnet strenghts change the dispersion function. The action of the L-FB due to these dispersive offsets has hardly any impact on the luminosity results. Furthermore, it can be observed that the tolerances for the quadrupoles in the final doublet are orders of magnitude tighter than for the quadrupoles in the main linac. While for the main linac a tolerance of $5\times 10^{-2}\,\%$ is necessary to keep the luminosity loss in the order of $0.1\,\%$, the final doublet quadrupoles have a tolerance of $10^{-4}\,\%$ to reach the same performance. The magnets of the final focus (without final doublet) and the BDS (without final focus) show tolerances between these values for the final doublet and the main linac.

Another important imperfection class are scaling errors of the BPM measurements and the corrector actuations. For the simulation of the BPM scaling errors the measurements (including BPM noise) were multiplied with the factor $1+a_i$, where a_i is a Gaussian distributed random number with standard deviation σ. The factor a_i is different for each BPM, but stays constant for all time steps of the simulation. The orbit response matrix used by the controller was assumed to be perfectly known. The results of the simulations are shown in Fig. 4.10 (left), where different seeds of ground motion model B as well as for model B10 have been investigated. For the main linac a scaling error of up to $100\,\%$ does not lead to a significant performance reduction, while the tolerance including the BDS is only about $1\,\%$ for a luminosity loss of $0.5\,\%$. The conjecture was made that these relatively tight tolerances originate mainly from the dispersive orbit in the horizontal direction. The measurement of the dispersive orbit could lead to large absolute measurement errors since the BPMs are not centred at the dispersive orbit but at zero. The black curve in Fig. 4.10 (left), for which only scaling errors in the horizontal

4.2. Imperfections and robustness

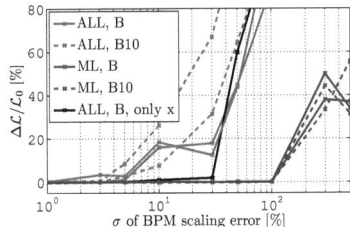 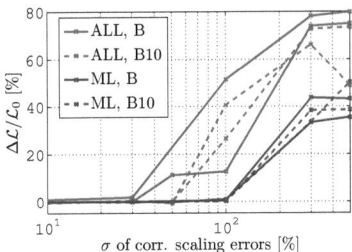

Figure 4.10.: Relative luminosity loss $\Delta\mathcal{L}/\mathcal{L}_0$ due to Gaussian distributed scaling errors of the BPMs (left) and correctors (right). Ground motion according to model B and B10 have been applied. Curves of equal colour and type correspond to different ground motion generator seeds for the same configuration of used ground motion model and scaling errors. The luminosity loss only due to ground motion (no scaling error) has been subtracted from the depicted curves.

direction are applied, shows however that the scaling error in the horizontal direction only has a minor effect on the overall luminosity loss. Therefore, the low tolerance of only 1 % originates from the high sensitivity of the L-FB to the vertical BPM scaling errors in the BDS.

In the same fashion as for BPM scaling errors, also the effect of corrector scaling errors has been investigated (see Fig. 4.10 (right)). Similar to the BPM scaling errors, no strong dependence of the luminosity loss to the used ground motion model was found. The tolerances for a luminosity loss of about 0.5 % are 100 % scaling error for only the main linac, and 30 % when also the BDS is included. These very high tolerances can be explained by considering that the corrector actuations of the L-FB are in the order of 0.1 nm. Scaling errors of the correctors consequently also create only a small error in the position of the quadrupoles, which explains the simulation results.

It should be mentioned that the effect of the scaling errors of the actuators has been only analysed with respect to positioning errors of the L-FB actuations. Another effect, which has been excluded from the studies, is that a scaling error of the quadrupole stabilisation system also causes differences in the ground motion suppression from quadrupole to quadrupole. This effect introduces a differential motion between the quadrupoles, which could lead to additional luminosity decrease. Studies of this effect are subject to future work.

Another imperfection, whose effect could be worsened by the L-FB, is the initial beam jitter. Kicks from the beam lines in front of the the main linac cause beam oscillations that correspond to beam position and angle offset at the entrance of the main linac. Realistic beam jitter can be created by the following procedure.

4. Controller performance, imperfections and robustness studies

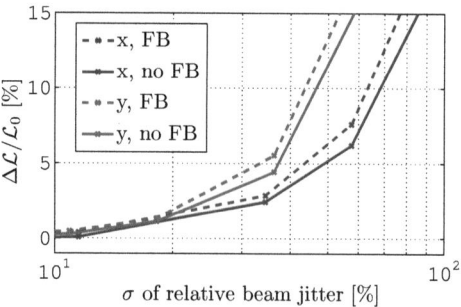

Figure 4.11.: Relative luminosity loss $\Delta\mathcal{L}/\mathcal{L}_0$ due to white, Gaussian beam offset jitter at the entrance of the main linac with a relative standard deviation σ_j.

The beam offset y_0 and angle y'_0 at the entrance of the main linac can be written as

$$y_0 = \sqrt{2J\beta_0}\cos(\Delta\phi) \quad \text{and} \tag{4.3}$$

$$y'_0 = -\sqrt{\frac{2J}{\beta_0}}\left[\sin(\Delta\phi) + \alpha_0\cos(\Delta\phi)\right], \tag{4.4}$$

where α_0 and β_0 are the twiss parameters at the entrance of the main linac, J is the action of the beam oscillation and $\Delta\phi$ is the phase advance between the origin of the kick and the entrance of the main linac. The phase advance $\Delta\phi$ is assumed to be a uniformly distributed random variable between $-\pi$ and $+\pi$, since the kicks in the beam line before the main linac occur in a random fashion. To create oscillations with a beam offset of a certain standard deviation σ_{y_0}, the action has to be chosen according to

$$\sigma_{y_0}^2 = \mathbb{E}\left\{y_0^2\right\} = \mathbb{E}\left\{2J\beta_0\cos^2(\Delta\phi)\right\} = J\beta_0, \quad \Rightarrow \quad J = \frac{\sigma_{y_0}^2}{\beta_0}. \tag{4.5}$$

Hence, realistic beam jitter can be created by first choosing J according to Eq. (4.5) and $\Delta\phi$ to be a uniformly distributed random number between $-\pi$ and $+\pi$. As a second step, the beam offset y_0 and beam angle y'_0 are calculated according to the Eqs. (4.3) and (4.4). Using this procedure, the luminosity loss due to initial beam jitter was determined via simulations. The according results are presented in Fig. 4.11 (left). It can be observed that for both, the vertical and the horizontal direction, the beam position jitter has to stay below 10% of the beam size to cause a luminosity loss of smaller 0.5%. The horizontal and the vertical direction show similar behaviour, with the vertical direction being slightly more sensitive. The action of the L-FB has only a small effect on the results.

4.3. Conclusions

As the main result, it can be stated that the luminosity loss due to ground motion can be reduced by the ground motion mitigation methods of CLIC to a level of about 12%

4.3. Conclusions

for model B10. This value already includes the effect of the limited resolution of the BPMs that accounts for about 2% of the loss. Due to the robustness of the L-FB the tolerances for the BPM resolution could be relaxed to 50 nm in the BDS and 100 nm in main linac. This low sensitivity to the BPM resolution (compared to older designs) can be achieved due to the optimised time-dependent and spatial filters of the L-FB presented in Chap. 3.

Another important outcome of the conducted simulations were guidelines for an improvement of the shape of the frequency response of the quadrupole stabilisation system. Artificial modification of the frequency response of the quadrupole stabilisation version 1 showed that a shift of the amplification peaks towards higher frequencies leads to an improvement of the luminosity performance. The quadrupole stabilisation group considered these guidelines and designed a new quadrupole stabilisation system named version 2, which is based on a different sensor than version 1 but uses the same tripod positioning system. The quadrupole stabilisation version 2 is currently under development, but performance predictions with an optimised L-FB show a luminosity loss as small as 0.8% for model B10.

Furthermore, the impacts of machine imperfections have been investigated. It turned out that the most severe problems are caused by beam energy errors produced by initial beam energy jitter or jitter in the acceleration gradients. These beam energy deviations cause large beam offsets in the collimation section of the BDS. The L-FB reacts on these large offsets, which results in a misteering of the beam and consequently in a large luminosity loss. A strategy, called dispersion filter, was developed to filter the energy dependent beam offsets from the BPM measurements, which resolves the problem. It was also shown that the dispersion filter can be used to measure the beam energy deviations with high accuracy. Another observation is that the tolerances for BPM scaling errors and positioning errors of the quadrupole stabilisation system are tight but feasible.

Finally, we want to lists the most important dynamic imperfections and the according luminosity losses in Tab. 4.2. Including all dynamic imperfections, the estimated luminosity loss is about 20%. Since by design a luminosity loss for dynamic imperfections of about 20% has been taken into account, the tight luminosity specifications of CLIC seem to be achievable.

4. Controller performance, imperfections and robustness studies

Imperfection	Expected value	$\Delta\mathcal{L}/\mathcal{L}_0$
Ground motion	model B10	9.7 %
BPM resolution for L-FB	$\sigma_{BPM} = 50\,\text{nm}$	2.0 %
BPM resolution for IP-FB	$\sigma_{BPM,IP} = 3\,\mu\text{m}$	0.1 %
RF jitter	combined	4 %
Initial beam energy jitter	$\sigma_{E_i} = 1\,\%$	0.1 %
Sensor noise of the quadrupole stabilisation V1		0.3 %
Quadrupole stabilisation system positioning errors	$\sigma_u = 0.25\,\text{nm}$	0.5 %
Quadrupole strength jitter	combined	0.5 %
BPM scaling error	$\sigma_{s,BPM} = 1\,\%$	0.5 %
Corrector scaling error	$\sigma_{s,corr} = 0.1\,\%$	0.0 %
Initial beam jitter	$\sigma_{x_i} = \sigma_{y_i} = 10\,\%$	0.7 %
BPM noise in the measurements of $\boldsymbol{R}$	$\sigma_n = 10\,\text{nm}$	2 %
Sum		20.4 %

Table 4.2.: Overview of the relative luminosity loss $\Delta\mathcal{L}/\mathcal{L}_0$ due to different dynamic imperfections, where all ground motion mitigation methods have been considered. For the given estimates, typical expected values of the imperfections have been used.

5. System identification scheme for orbit response matrices

In this chapter, we present an on-line system identification scheme that is capable of measuring the orbit response matrix R and improving the quality of the measurement over time, without stopping the data tacking of the detectors. Furthermore, changes of R over time can be learned by the algorithm, which is based on methods from the field of system identification theory. The algorithm is optimised for the main linac of CLIC, which is an especially difficult system to identify. The measured orbit response matrix can be used to ensure good performance of the L-FB over time and as an input for diagnosis tools. Most of the material presented in this chapter was published also in Pfingstner et al. [85] and Pfingstner et al. [84].

A general introduction to the on-line system identification scheme for orbit response matrices is given in Sec. 5.1. More detailed information about the algorithm are presented in Sec. 5.2 and App. C.5. For the identification of the main linac of CLIC, the basic algorithm has to be extended with the help of an amplitude model of the beam oscillations in the main linac, which is presented in Sec. 5.3. The combination of the on-line identification algorithm and the L-FB is covered together with simulation results in Sec. 5.4. Finally, conclusions are given in Sec. 5.5.

5.1. Introduction

5.1.1. Motivation

Good system knowledge is an essential ingredient for the successful operation of modern particle accelerators. For the L-FB, beam-based alignment as well as for diagnosis and error detection methods, the most important system information is the orbit response matrix R. Factors that limit the measurement accuracy of R are the influence of measurement noise, ground motion and beam energy variations during the measurement process.

Due to the importance of the quality of the measured orbit response matrix, we present in this chapter an algorithm that is capable of improving the measurement accuracy of R over time. Furthermore, the algorithm is capable of learning time-dependent changes of R on-line, without stopping physics data taking. To accomplish these tasks, methods from the field of system identification are used, which are adapted to the special structure of the accelerator environment. The algorithm is tested via simulations on the main linac of CLIC. The extension of the algorithm to the BDS is subject to future work.

The orbit response matrix estimated by the system identification algorithm can be used for multiple purposes. The performance of the L-FB can be improved by updating the orbit response matrix used by the L-FB with the estimated $\hat{R}[k]$ from the system

5. System identification scheme for orbit response matrices

identification algorithm, where k is the time index. This procedure corresponds to an adaptive control algorithm or more precisely to a *self-tuning regulator* (STR). Other important applications for the estimated $\hat{R}[k]$ are diagnosis tools. Tools similar to LOCO (see Safranek [105]) can be used to determine the variation of certain accelerator parameters, as e.g. acceleration gradients, from the change of the orbit response matrix. Also breakdowns and scaling errors of BPMs and correctors can be detected.

5.1.2. System identification for orbit response matrices

The field of system identification is concerned with the determination of unknown parameters of a system from measurements. The structure of the system to be identified is assumed to be known in the approach used in this thesis (white box identification). To be able to estimate the unknown parameters, an excitation unit produces excitations $u[k]$ that are applied to the system to be identified. These excitations as well as the outputs $y[k]$ of the system are used by an estimation algorithm to create an estimate of the unknown parameters. The task of the estimation algorithm is complicated by the fact that additionally to the known $u[k]$, also unknown disturbances $d[k]$ (ground motion) are present at the inputs of the systems. Furthermore, the measurements of the output values contain measurement noise $n[k]$. The overall structure of the system is shown in Fig. 5.1 (left).

In this thesis, we use as an estimation algorithm the *recursive least squares algorithm (RLS) with exponential forgetting*. The algorithm is described in App. C.5, but we want to comment on the expression "exponential forgetting" at this position. If the estimation algorithm would weight all measurement in the same way, new measurements would have hardly any influence on the estimation result after long estimation periods. Changes of the system behaviour would be learned slower and slower with increasing estimation time. Instead, the algorithm has to "forget" older values, by applying a smaller weight to them. In the case of the used algorithm, this weighting is performed with an exponential function with a time constant α called *forgetting factor*.

In the case of the main linac of CLIC the system to be identified has the form

$$y[k] = R[k]\left(u[k-1] + d[k]\right) + n[k], \tag{5.1}$$

where $R[k]$ is the time-varying orbit response matrix of the main linac. A description of the properties and the shape of the orbit response matrix has already been given in Sec. 3.2.1. We only want to restate here that the ith column of $R[k]$ corresponds to the beam oscillation measured in all used BPMs due to an excitation of the ith corrector (in this case the ith quadrupole) with an unit step of $1\,\mu$m. An example for such an oscillation is given in Fig. 5.2 (left).

There are two problems arising when we want to identify $R[k]$ on-line. The first is the necessary identification time. To be able to identify all elements of $R[k]$, which has the dimensions 2010×2010, the excitation has to be persistent (for a definition of persistent excitation see Åström and Wittenmark[95]). For the system Eq. (5.1) this is equivalent to applying at least 2010 linear independent input vectors. With a beam repetition rate of 50 Hz, one full identification cycle would take 40.2 s. Many of such cycles are necessary to create an accurate estimate of R (number depends on the needed accuracy,

5.1. Introduction

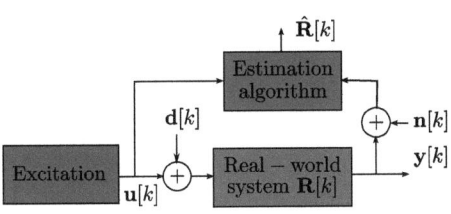 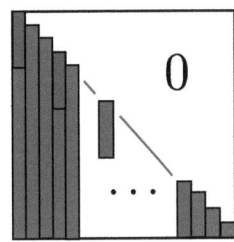

Figure 5.1.: (left) Principle of a system identification algorithm. A system identification algorithm consists of an excitation unit and an estimation algorithm (in red) that produce an estimate $\hat{R}[k]$ of the real system behaviour $R[k]$ (in blue). The excitations $u[k]$ are applied to the real-world system that create together with the unknown disturbances $d[k]$ the system outputs $y[k]$. The measurements $y[k] + n[k]$, where $n[k]$ is the measurement noise vector, and the excitations $u[k]$ are used by the estimation algorithm to produce the system estimate $\hat{R}[k]$. (right) The columns of the orbit response matrix $R[k]$ are composed out of the beam oscillations due to kicks from the different correctors. The matrix $R[k]$ is triangular, since the beam oscillations only result in offsets in BPMs downstream of the used corrector. Instead of estimating all elements of R, only a subset of the parameters (marked in red) is identified but by the independently running RLS algorithms.

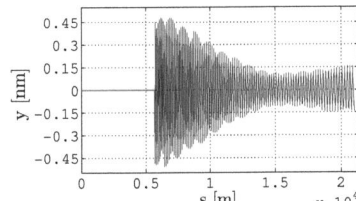 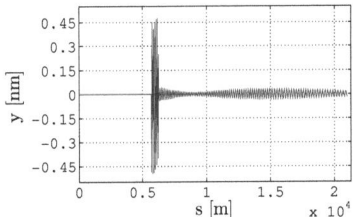

Figure 5.2.: (left) Vertical beam oscillation along the main linac of CLIC due to a corrector kick at position $5678\,m$. The beam motion can be decomposed into an amplitude $A(s)$ and a phase $\phi(s)$. The shape of the amplitude is determined by the quadrupoles and filamentation (see Sec. 5.3). The waist at $16\,km$ is a typical effect of the filamentation. (right) Orbit bump number 34 with a starting location of $5678\,m$. The beam is excited due to a kick u_{c1}. Shortly after, it is steered onto its original orbit, by applying two other kicks u_{c2} and u_{c3}. The bump closure is not perfect as can be seen from the remaining beam motion after the bump.

5. System identification scheme for orbit response matrices

the allowable emittance growth and the noise level). The algorithm is therefore only useful to identify very slow variations.

The second problem is even more severe. If the excited beam would oscillate all the way through the accelerator, the emittance would increase in a not acceptable way. Smaller excitations would cause longer identification times in order to compensate the lower signal to noise ratio.

5.1.3. Interleaved, model-supported system identification scheme

To decrease the necessary identification time, we propose the following scheme. Only a subset of all correctors is used to excite the beam. Theses correctors are not used at the same time step, but actuated one after the other. Therefore, the resulting beam oscillations can be observed independently. To allow for a large beam oscillation without increasing the multi-pulse emittance too strongly, the beam is excited to oscillate just over a short distance in the main linac and is kicked back afterwards onto its original orbit. Such local beam position changes are called orbit bumps (see Fig. 5.2 (right)). The lengths and amplitudes of the orbit bumps are chosen in a way, such that the caused multi-pulse emittance increase is acceptable (< 0.1 nm rad). The calculation of the excitations that create such bumps and the used setup will be explained in Sec. 5.2.1. Since not every quadrupole is used to create an orbit bump the identification time is strongly reduced. Each of the orbit bump excitations can be used to locally identify elements of $R[k]$ by applying for every orbit bump one RLS algorithm with exponential forgetting. Due to the special structure of the system Eq. (5.1) the usually computationally expensive RLS algorithm can be calculated very efficient. This will be shown in Sec. 5.2.2. The disadvantage of this scheme is that only the starting elements of some columns of $R[k]$ can be identified (marked in red in Fig. 5.1 (right)).

In order to combine these local identification results to a complete orbit response matrix, priori knowledge has to be used. Each column of $R[k]$ corresponds to an oscillation with varying phase and amplitude. The phase advance ϕ between different locations $s_A < s_B < s_C$ in the main linac is additive, i.e. $\phi_{AC} = \phi_{AB} + \phi_{BC}$. The total phase advance for the whole main linac and therefore for every column of $R[k]$ can be reconstructed by combining the phase information of all local RLS results, since the bumps have been positioned in an interleaved fashion. The calculation and the merging of the local phase information is presented in Sec. 5.2.3.

The missing piece to reconstruct $R[k]$ is the amplitude information. Since the amplitude shape shows just small changes due to accelerator component variations, its form can be modelled and assumed to be constant with time. The according amplitude model is derived in Sec. 5.3. The absolute amplitude of this model is scaled to the proper size using the amplitudes of the identification results. The combination of amplitude and phase information allows to produce an estimate $\hat{R}[k]$ of the orbit response matrix $R[k]$. The described procedure is visualised in Fig. 5.3 in a simplified manner.

When reconstructing $\hat{R}[k]$, it has to be taken into account that the beam oscillation amplitudes are different for a focusing or defocusing quadrupole (two different types of quadruples that are used in pairs in the main linac) as an actuating element. This difference is only a scaling factor that does not change the shape of the amplitudes along the linac. A corresponding scaling factor between the oscillation amplitude for a

5.1. Introduction

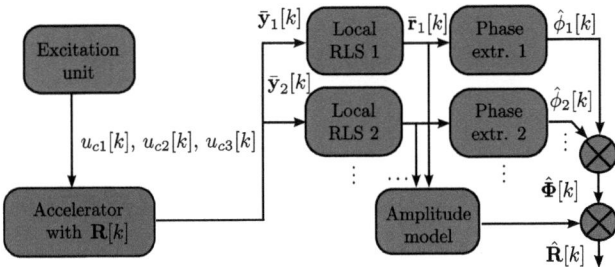

Figure 5.3.: Block diagram of the proposed identification scheme. The excitation unit excites the beam locally by sequentially creating orbit bumps with corrector actuations $u_{c1}[k]$, $u_{c2}[k]$ and $u_{c3}[k]$ along the accelerator. The according measurement data $\bar{y}_i[k]$ are used to identify parts of certain columns of $R[k]$, named $\bar{r}_i[k]$, with the help of local RLS algorithms. The identified data are used to extract phase and scaling information. The overall phase $\hat{\Phi}[k]$ is reconstructed by combining the local phases $\hat{\phi}_i[k]$. The estimated response matrix $\hat{R}[k]$ is filled by combining the phase information and the scaled amplitude model.

unit excitation with a focusing and defocusing quadrupole is estimated by the system identification algorithm along the main linac. Similarly, also a factor (as a function of the position in the linac) between the oscillation amplitude in BPMs corresponding to focusing and defocusing quadrupoles is estimated and used for the matrix reconstruction.

5.1.4. Related work

There are many good references, for a general introduction to the field of system identification and the RLS algorithm, e.g. Åström and Wittenmark [95] and Ljung and Gunnardsson[70]. However, concerning applications in the field of particle accelerators and especially for orbit response matrices and also adaptive orbit feedbacks, the literature is rather limited. Therefore, we mention here as well work that used system identification only indirectly, as e.g. adaptive control systems.

Barr [10] describes the use of a self-tuning regulator (STR) to adapt parameters of several local, independent orbit feedback loops, which are distributed along the accelerator. This is obviously a sub-optimal approach, since the mutual interaction of the different feedback loops is not taken into account. Himel [56] goes one step further. Instead of using independent loops, each loop corrects just the errors created between itself and the loop before. This is accomplished by using the beam measurements of the loop upstream and an estimate of the propagation of these parameters to the actual loop. The estimate of the propagation is created by identification of the beam transfer matrix between the bumps. However, also this technique is sub-optimal due the local nature of the feedback system. The system identification scheme present in this chapter will allow us to establish a global estimate of the response matrix. Such an estimate can be used by a global feedback algorithm which is best suited to mitigate ground motion effects.

5. System identification scheme for orbit response matrices

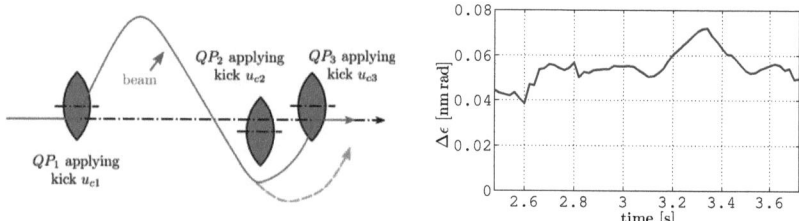

Figure 5.4.: (left) Scheme of an orbit bump. The first displaced quadrupole kicks the beam with a certain strength u_{c1}. The kick u_{c2}, from the second quadrupole, changes the trajectory of the beam in a way that it arrives in the third quadrupole with no vertical offset. The third quadrupole can now compensate the vertical velocity component of the beam by applying a proper third kick u_{c3}. Hence, the beam continues after the third quadrupole on its initial orbit. (right) The (normalise, multi-pulse) emittance growth over one cycle of identification. Since the multi-pulse emittance is an averaged value, it has a certain transient behaviour. Therefore the plot shows the third and not the first cycle.

5.2. Details about the identification scheme

In the following, more details about the proposed system identification scheme are given. The excitation unit, the local RLS algorithm and the phase reconstruction are described. The also necessary amplitude model is covered separately in Sec. 5.3.

5.2.1. Excitation unit

The excitation unit excites the beam with so called orbit bumps. To create such an orbit bump, at least three quadrupoles are necessary (see Wille [136]). The scheme is visualised in Fig. 5.4 (left). Even thought an orbit bump can in principle steer the beam after the bump perfectly onto its initial orbit (bump closure), the necessary precise system knowledge is not available (R is not enough since also information about the angle of the beam are needed). A sub-optimal approach is used. The first quadrupole QP_1 excites the beam to an oscillation of proper amplitude. The kicks u_{c2} and u_{c3} of the second and third quadrupole QP_2 and QP_3 are chosen, such that the measured amplitude y_b in all BPMs after the orbit bump is minimised in a quadratic sense. In the current setting QP_2 is located directly in front of QP_3, while between QP_1 and QP_2 there are several other quadrupoles that are not listed here to create a appropriate bump length. To calculate u_{c2} and u_{c3} we use the relationship

$$y_b = r_{c1}u_{c1} + r_{c2}u_{c2} + r_{c3}u_{c3}, \qquad (5.2)$$

where r_{c1}, r_{c2} and r_{c3} are the elements of the c_1^{th}, c_2^{th} and c_3^{th} column of $R[k]$. The amplitudes y_b can be minimised in a quadratic sense by choosing u_{c2} and u_{c3} as

$$\begin{bmatrix} u_{c2} \\ u_{c3} \end{bmatrix} = -\begin{bmatrix} r_{c2} & r_{c3} \end{bmatrix}^\dagger r_{c1}u_{c1}, \qquad (5.3)$$

5.2. Details about the identification scheme

where † stands for the pseudo-inverse of a matrix. The orbit bumps created in this manner are not closed perfectly as can be seen in Fig. 5.2 (right), since only two correctors are not zero all BPM readings perfectly.

For the identification algorithm overlapping orbit bumps are created along the accelerator. The kick strength u_{c1} of each bump is optimised to create beam oscillations with a maximal amplitude of about 0.4 to 0.5 µm. The length of the bump is chosen such that the emittance growth stays below 0.1 nm rad. The therefor necessary bump length was determined via simulations. It varies along the main linac and also depends if a focusing or a defocusing quadrupole is used for the excitation. Each bump overlaps with the bump before by about 28 BPMs on average to ensure an accurate phase merging (see Sec. 5.2.3). Altogether, 62 orbit bumps have been created in this way along the main linac. The emittance growth cause by these 62 excitations (one identification cycle) is depicted in Fig. 5.4 (right).

Since just 62 beam bumps are created, the time needed for one complete identification cycle is strongly reduced, compared to a conventional RLS algorithm, for which 2010 linear independent excitations would be applied for one identification cycle. With a beam repetition rate of 20 ms, this results in an identification cycle time of 1.24 s instead of 40.2 s. This corresponds to an improvement in identification speed by a factor of about 33. Notice however, that the excitation in our algorithm is not persistent anymore.

5.2.2. Local RLS algorithm

In this section, we assume that the reader is familiar with the RLS algorithm with exponential forgetting. If this is not the case, a brief introduction is given in App. C.5 and more detailed information can be found in Åström and Wittenmark [95] and Ljung and Gunnardsson [70]. The general RLS algorithm with exponential forgetting is given by

$$\hat{\boldsymbol{\theta}}[k] = \hat{\boldsymbol{\theta}}[k-1] + \boldsymbol{K}[k](y[k] - \boldsymbol{\varphi}[k]^T\hat{\boldsymbol{\theta}}[k-1]) = \hat{\boldsymbol{\theta}}[k-1] + \boldsymbol{K}[k]e[k] \qquad (5.4)$$

$$\boldsymbol{K}[k] = \boldsymbol{P}[k-1]\boldsymbol{\varphi}[k](\lambda\boldsymbol{I} + \boldsymbol{\varphi}[k]^T\boldsymbol{P}[k-1]\boldsymbol{\varphi}[k])^{-1} \qquad (5.5)$$

$$\boldsymbol{P}[k] = (\boldsymbol{I} - \boldsymbol{K}[k]\boldsymbol{\varphi}[k]^T)\boldsymbol{P}[k-1]/\lambda, \qquad (5.6)$$

where k represents the time index, $\hat{\boldsymbol{\theta}}[k]$ contains the estimated system parameters, $y[k]$ are the measurement values, $\boldsymbol{K}[k]$ contains error weighting factors, $\boldsymbol{\varphi}[k]^T$ the excitation values, $e[k] = y[k] - \boldsymbol{\varphi}[k]^T\hat{\boldsymbol{\theta}}[k-1]$ are the errors between the estimated and real output data, $\boldsymbol{P}[k]$ is the a matrix measuring the accumulated excitation strength and $\boldsymbol{I}$ is the identity matrix.

In general, the RLS algorithm with exponential forgetting is computational expensive. Therefore, simplified algorithms are available as e.g. the *stochastic approximation algorithm* (SA) and the least mean square algorithm (LMS). However, the simple structure of the accelerator system in Eq. (5.1), can be exploited to reduce the computational effort significantly. As a result the full RLS algorithm can be used. The usually necessary matrix inversion reduces to a simple scalar division. Additionally, just four numbers per row of $\boldsymbol{R}$ have to be stored to reconstruct $\boldsymbol{P}[k]$. These statements will be shown in the following derivations.

5. System identification scheme for orbit response matrices

Due to the local excitation, the complete systems Eq. (5.1) reduces for one orbit bump to

$$\bar{y}[k] = \bar{r}[k]u_{c1}[k] + \bar{y}_d[k] + \bar{n}[k], \qquad (5.7)$$

where $\bar{y}[k]$ are the BPM measurements at locations where the beam is excited by the bump (between QP_1 and QP_2), $\bar{r}[k]$ are the according elements of $\boldsymbol{R}$ excited by the beam bump, $\bar{y}_d[k]$ are constant beam offsets due to a non-zero reference orbit, slow drifting BPM values due to ground motion effects or other slow changing imperfections and $\bar{n}[k]$ is white Gaussian measurement noise with a standard deviation of 50 nm. Applying the general RLS algorithm in Eqs. (5.4), (5.5) and (5.6) to the system Eq. (5.7) results in the following expressions for $e[k]$, $P[k]$, $K[k]$ and $\hat{\boldsymbol{\theta}}[k]$ for the local RLS algorithms.

$$e[k] = \bar{y}[k] - \bar{\varphi}[k]^T \hat{\boldsymbol{\theta}}[k-1]$$

$$= \begin{bmatrix} \bar{y}_1[k] \\ \bar{y}_2[k] \\ \vdots \\ \bar{y}_m[k] \end{bmatrix} - \begin{bmatrix} [u_{c1}[k]\ 1] & & 0 \\ & \ddots & \\ 0 & & [u_{c1}[k]\ 1] \end{bmatrix} \begin{bmatrix} \bar{r}_1[k-1] \\ \bar{y}_{d,1}[k-1] \\ \bar{r}_2[k-1] \\ \bar{y}_{d,2}[k-1] \\ \vdots \\ \bar{r}_m[k-1] \\ \bar{y}_{d,m}[k-1] \end{bmatrix}$$

$$= \bar{y}[k] - \bar{r}[k-1]u_{c1}[k] - \bar{y}_d[k-1],$$

For the following derivation, it will be useful to show that $P[k]$ has a diagonal form, if the compound matrix notation is used. For this reason, we use a relationship taken from Åström and Wittenmark [95] for the RLS without forgetting factor to get

$$P[k] = \left(\sum_{i=1}^{k} \bar{\varphi}[k]\bar{\varphi}[k]^T\right)^{-1} = \begin{bmatrix} \bar{\Phi}[k]^{-1} & \cdots & 0 \\ \vdots & \ddots & \vdots \\ 0 & \cdots & \bar{\Phi}[k]^{-1} \end{bmatrix} \text{ with} \qquad (5.8)$$

$$\bar{\Phi}[k] = \sum_{i=1}^{k} \begin{bmatrix} u_{c1}[i]^2 & u_{c1}[i] \\ u_{c1}[i] & 1 \end{bmatrix}. \qquad (5.9)$$

The fact that $P[k]$ has a diagonal structure is also true for the RLS algorithm with exponential forgetting, since the older measurement values are just degraded over time and we assume a diagonal form of the initial covariance matrix. Using this simplification we further get

$$K[k] = P[k-1]\bar{\varphi}[k](\lambda I + \bar{\varphi}[k]^T P[k-1]\bar{\varphi}[k])^{-1}$$

$$= \begin{bmatrix} \bar{P}[k-1] & \cdots & 0 \\ \vdots & \ddots & \vdots \\ 0 & \cdots & \bar{P}[k-1] \end{bmatrix} \begin{bmatrix} \begin{bmatrix} u_{c1}[k] \\ 1 \end{bmatrix} & \cdots & 0 \\ \vdots & \ddots & \vdots \\ 0 & \cdots & \begin{bmatrix} u_{c1}[k] \\ 1 \end{bmatrix} \end{bmatrix} \begin{bmatrix} b[k] & \cdots & 0 \\ \vdots & \ddots & \vdots \\ 0 & \cdots & b[k] \end{bmatrix}$$

5.2. Details about the identification scheme

$$\boldsymbol{K}[k] = \begin{bmatrix} \bar{\boldsymbol{k}}[k] & \cdots & 0 \\ \vdots & \ddots & \vdots \\ 0 & \cdots & \bar{\boldsymbol{k}}[k] \end{bmatrix} \quad \text{with} \tag{5.10}$$

$$b[k] = \cfrac{1}{\lambda + \begin{bmatrix} u_{c1}[k] & 1 \end{bmatrix} \bar{\boldsymbol{P}}[k-1] \begin{bmatrix} u_{c1}[k] \\ 1 \end{bmatrix}} \quad \text{and} \tag{5.11}$$

$$\bar{\boldsymbol{k}}[k] = \begin{bmatrix} \bar{k}_1[k] \\ \bar{k}_2[k] \end{bmatrix} = b[k]\bar{\boldsymbol{P}}[k-1]\begin{bmatrix} u_{c1}[k] \\ 1 \end{bmatrix}. \tag{5.12}$$

In a similar way $\boldsymbol{P}[k]$ can be shown to be of the form

$$\boldsymbol{P}[k] = \begin{bmatrix} \bar{\boldsymbol{P}}[k] & \cdots & 0 \\ \vdots & \ddots & \vdots \\ 0 & \cdots & \bar{\boldsymbol{P}}[k] \end{bmatrix} \quad \text{with} \tag{5.13}$$

$$\bar{\boldsymbol{P}}[k] = \begin{bmatrix} 1 - \bar{k}_1[k]u_{c1}[k] & -\bar{k}_1[k] \\ -\bar{k}_2[k]u_{c1}[k] & 1 - \bar{k}_2[k] \end{bmatrix} \bar{\boldsymbol{P}}[k-1]/\lambda. \tag{5.14}$$

Using the structure of $e[k]$ and $\boldsymbol{K}[k]$ the coefficient update $\hat{\boldsymbol{\theta}}[k]$ can be written as

$$\bar{r}[k] = \bar{r}[k-1] + \bar{k}_1[k]e[k] \quad \text{and} \tag{5.15}$$
$$\bar{y}_d[k] = \bar{y}_d[k-1] + \bar{k}_2[k]e[k]. \tag{5.16}$$

5.2.3. Phase reconstruction and merging

The data identified by the local RLS algorithms represent beam oscillations along the main linac. Hence, the identification results can be used to extract the phase information of the beam oscillations of the according section of the main linac. The algorithm performing this task uses the following strategy. First, the zero crossings in the data are detected. A zero crossing between two consecutive BPMs has happened, if the according measurements have different signs. The position of the zero is estimated by determining the zero crossing of the linear interpolation between these two measurements, since the BPM positions are close to each other compared to the beam oscillation wave length. To become more robust against measurement noise, an additional check is performed. It uses the knowledge that the phase advance is smooth along the main linac. If a zero is unusually close to another zero, it is most likely just a noise artefact. The zero deviating more from the expected position can be deleted.

With the detected zero crossings the phase advance for each bump can be reconstructed. The phase advance for every bump starts with $0°$ at the position of the first quadrupole. This is due to the fact that the beam is kicked out of its nominal orbit into a sine-like oscillation. At each zero crossing the phase advance has grown by $180°$ compared to the zero crossing before. Since the phase advance can be assumed to be smooth, the phase advance between zero crossings is determined by linear interpolation/extrapolation. To reduce the local phase error, introduced by the zero detection, the generated phase data are low-pass filtered in space domain, by applying a Hamming window.

5. System identification scheme for orbit response matrices

After the local phase information have been extracted, they can be merged to one global phase by the following technique. Since the bumps are chosen to be overlapping, also the phase information of two consecutive bumps are overlapping. The phase information of a bump is stitched to the one before, by simply adding a proper offset value. This offset value is chosen in a way that the mean value of the error between the two phase information in the overlapping region is minimised. It should be mentioned that the assumption of an additive phase is not completely valid. For the generation of $R[k]$ the phase has to be slightly modified to achieve the necessary accuracy. However, this is an accumulated and static effect, which is not significantly changing with varying accelerator component drifts.

5.3. Amplitude model for the main linac of CLIC

5.3.1. Introduction

The vertical beam motion in the main linac due to an excitation (kick) is sinusoidal-like. It can be described by an amplitude $Y(s)$ and a phase advance $\phi(s)$, where s is the position in the main linac in metre (see Fig. 5.2 (left)). This section focuses on a model for $Y(s)$. It is determined by the so called beta function of the accelerator and a particle-energy dependent effect called filamentation. Even tough the principle of filamentation will just be explained in the next section, we already state here the connection to the relevant literature.

In the literature, filamentation is mentioned in connection with the term of Landau damping. Landau damping is the damping of certain beam instabilities due to filamentation. Even though we are not interested in these damping effects, the literature about Landau damping provides the basic formalism for our studies. However, this basic formalism has to be modified, as for example Chao [22] and Hofmann [58] simplify the focusing strength of the accelerator components to a constant value and study mainly beam stability in rings, which is not a valid simplification for the main linac of CLIC. In the mentioned literature, the observed filamentation is also just observed as time goes to infinity. This is also not a useful treatment for the current application, where the transient effects due to the special accelerator design are the important ones.

In this section, we will therefore adapt the general model to quantify filamentation and calculate $Y(s)$. We will present approximations of the beta function of the main linac of CLIC and the particle motion dependence on the energy and the particle energy distribution function in order to find an accurate expression for $Y(s)$. At positions in the main linac, where made assumptions are not valid, the basic model will be extended with a fit to simulation data.

5.3.2. Description of beam oscillations in the main linac of CLIC

As already mentioned, the vertical beam position $Y(s)$ is determined by the so called beta function and the effect of filamentation. The beta function characterises the effects of the accelerator components, e.g. magnets and accelerating structures, on the beam. The beta function is determined by the accelerator design and fully describes the motion of a particle with given energy along the beam line. Filamentation, on the other hand,

5.3. Amplitude model for the main linac of CLIC

is a result of the fact that particles with different energies will be affected differently by the accelerator components. The vertical beam position $Y(s)$ in the main linac is the average of the positions of the individual particles $Y(s) = \mathbb{E}\{y(s)\}$. When the beam is kicked out of its nominal orbit, the particle motion due to the kick is in phase. However, particles with lower energy oscillate faster than particles with higher energy. Slowly a phase difference builds up between particles with different energies. The particles move less and less coherently. Due to this ensemble of sinusoidal waves with different wave lengths, the average beam oscillation will be damped. A characteristic for filamentation is the occurrence of waists and maxima in the beam oscillation amplitude. This is due to the fact that different particle oscillations add up at some positions along the accelerator in a coherent and at some positions in an incoherent way.

5.3.2.1. General approach

If the beam is in its nominal orbit, all particles of the beam move along the beam line with some amplitude and the betatron phase advance. Since the individual particles have random phases their average position (beam position) is zero. In case a certain kick $\Delta y'$ is applied to the beam at some position s_k, the whole beam behaves much like every individual particle. The kick superimposes a second motion (also the betatron motion) to the individual motion of every particle. Since this second motion is coherent for every particle (same phase for every particle), the beam as a whole starts to oscillate. The amplitude A_k of that motion will depend on the size and the position of the kick, and is assumed to be known. Hence $Y(s)$ can be written as

$$Y(s_k, s) = \begin{cases} 0, & 0 \leq s \leq s_k \\ A_k \int\limits_{-\infty}^{\infty} y(s_k, s, \delta) \rho(\delta) d\delta, & s_k < s \leq s_{\text{end}} \end{cases}, \quad (5.17)$$

$$(5.18)$$

where s_{end} is the length of the main linac of 21028 m, $\delta = (E_i - E_0)/E_0$ with $E_0 = \mathbb{E}\{E_i\}$ is the particle's energy deviation from the average particle energy, $y(s_k, s, \delta)$ is the motion of a particle with a certain δ and $\rho(\delta)$ is the distribution function of the particles due to their δ.

Recognise that the approach in Eq. (5.17) has to make the assumption that δ for a particle is constant along the main linac. This also implies that $\rho(\delta)$ does not change as a function of s. It is not obvious that this assumption can be made. For some parts of the main linac (entrance and exit) it is not even true. These parts will have to be treated separately, which will lead to extensions of the basic model.

However, for most parts of the system the constant δ assumption is valid (see Schulte [111]). The reason is a coherent energy spread (see Fig. 5.5) along the bunch, which exists in most parts of the main linac. Particles with different energies are separated longitudinally and can therefore be exposed to different acceleration gradients. Due to that fact, δ can be approximately constant for every particle. In the following, the general expressions $y(s_k, s, \delta)$ and $\rho(\delta)$ in Eq. (5.17) will be specialised for the main linac of CLIC.

5. System identification scheme for orbit response matrices

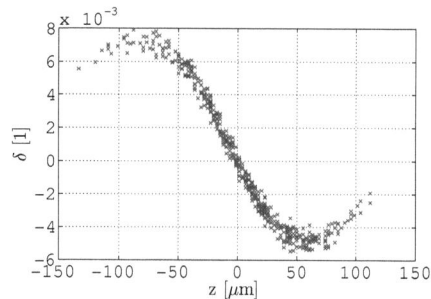

Figure 5.5.: Coherent energy spread $\delta = E_{part}/E_{avg}$ along the longitudinal dimension of one bunch (z) at position s = 6400 m in the main linac. A negative z corresponds to the front part of the bunch. The shape of the particles distribution develops due to the combined effects of acceleration voltage, wake field kicks and BNS damping (see Schulte [109] for more information).

5.3.2.2. Single particle motion with nominal energy

The motion of a single particle in the main linac with nominal energy ($\delta = 1$) is of the form

$$y(s_k, s) = A_k \frac{A(s)}{A(s_k)} \sin(\phi(s) - \phi(s_k)), \quad \text{for} \quad s_k < s \leq s_{\text{end}}. \tag{5.19}$$

For $0 \leq s \leq s_k$ the beam oscillations are $y(s_k, s)$ are zero. Notice that the amplitude of $y(s)$ is normalised to the size of A_k at the kick position. In principle $A(s)$ and $\phi(s)$ can be calculated by solving Hill's equation (see Wille [136]). However, for a complex beta function no closed solution can be found. Instead, the average beta function of the main linac was approximated by

$$\beta(s) = \sqrt{bs + c} \quad \text{with} \quad b = 0.051 \quad \text{and} \quad c = 4.8. \tag{5.20}$$

From this beta function, the beam size $\sigma(s)$ as well as the phase advance $\phi(s)$ of the single particle motion can be calculated as (see e.g. Wille [136])

$$\sigma(s) = \sqrt{\beta(s) \frac{\epsilon_N}{\gamma(s)}} \quad \text{and}$$

$$\phi(s) = \int_0^s \frac{1}{\beta(\acute{s})} d\acute{s} = \frac{2}{\sqrt{b}} \left(\sqrt{s + \frac{c}{b}} - \sqrt{\frac{c}{b}} \right), \tag{5.21}$$

where ϵ_N is the normalised emittance and $\gamma(s)$ the relativistic factor, which is the total particle energy divided by the rest mass energy. The beam oscillation amplitude behaves in the same way as the beam size. By stating the fact that γ grows with good approximation linear along the main linac and neglecting the absolute size of the oscillation by

5.3. Amplitude model for the main linac of CLIC

normalizing it, we can write $A(s)$ as

$$\frac{A(s)}{A(s_k)} = \frac{1}{\hat{A}(s_k)}\sqrt{\frac{\beta(s)}{\gamma(s)}} = \frac{1}{\hat{A}(s_k)}\sqrt{\frac{\sqrt{bs+c}}{ds+e}} \quad \text{with} \tag{5.22}$$

$$d = (1500-9)/(21 \times 0.51), \quad e = 9000/0.51 \quad \text{and} \quad \hat{A}(s_k) = \sqrt{\frac{\sqrt{bs_k+c}}{ds_k+e}}.$$

5.3.2.3. Single particle motion with non-nominal energy

Equations (5.21) and (5.22) approximate the motion of particles with nominal energy. To be able to describe the energy dependent effect of filamentation the expressions for $A(s)$ and $\phi(s)$ have to be generalised to $A(s,\delta)$ and $\phi(s,\delta)$. The dependence of A in respect to δ is relatively small (few percent, verified in Schulte et al. [113]). Since A appears as a multiplicative factor, these inaccuracies can be neglected.

Also the dependence of ϕ in respect to δ is small. However, since the values for ϕ becomes large, also small variations can change the according sine function of the beam oscillations completely. A derivation for the energy-dependent phase change in the main linac of CLIC can be found in Schulte [112]. This relative phase change for one FODO cell (elementary building blocks of the main linac consisting of one focusing and one defocusing magnet) is given by

$$\frac{\Delta\phi}{\phi_0} \approx -\frac{2}{\phi_0}\tan\left(\frac{\phi_0}{2}\right)\delta = -f\delta = -1.15\delta, \tag{5.23}$$

where $\phi_0 = 1.26$ is the phase advance of the FODO cells in the main linac of CLIC for nominal energy. Since the whole main linac consists of FODO cells, the complete main linac will have the same energy dependent behaviour as one FODO cell. Even though Eq. 5.23 delivers a good first approximation for the phase advance change, simulations have shown that small nonlinear effects appear. Especially a small asymmetry between different signs of δ was observed. Therefore, an average value of $f = -1.06$ was chosen for the following calculations. Using Eq. (5.23) to generalise Eq. (5.19) gives

$$y(s_k, s, \delta) = \begin{cases} 0, & 0 \leq s \leq s_k \\ A_k \dfrac{A(s)}{A(s_k)}\sin(\hat{\phi}(s)(1-f\delta)), & s_k < s \leq s_{\text{end}} \end{cases} \quad \text{with} \tag{5.24}$$

$$\hat{\phi}(s) = \phi(s) - \phi(s_k). \tag{5.25}$$

5.3.2.4. Particle energy distribution

As mentioned above, the main assumption of the current model is a constant energy distribution function $\rho(E)$ along the main linac. If different distributions along the main linac are plotted, one finds a typical distribution shown in Fig. 5.6 (right). This function can be approximated by the quadratic function

$$\rho(\delta) = \begin{cases} 0, & \delta < -0.0047 \\ \dfrac{g}{j}(\rho-h)^2 + \dfrac{i}{j}, & -0.0047 \leq \delta \leq 0.0066 \\ 0, & \delta > 0.0066 \end{cases} \quad \text{with} \tag{5.26}$$

5. System identification scheme for orbit response matrices

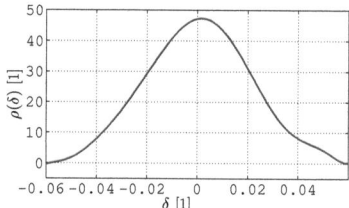
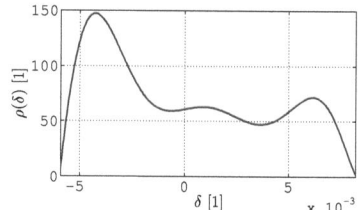

Figure 5.6.: Particle energy distribution ($\delta = (E_i - E_o)/E_o$ with $E_0 = \mathbb{E}\{E_i\}$) at position $s = 0$ m (left) and $s = 6400$ m (right) in the main linac of CLIC (9th order polynomial fit to simulation data). The Gaussian like shape of the plot on the left side is produced by the upstream transfer line and will be changed by the main linac gradually. At a s of 6400 m e.g., the shape of the curve is determined by the coherent longitudinal energy spread.

$$g = 3 \times 10^6, \quad h = 0.0025, \quad i = 50 \quad \text{and} \quad j = 1.0225,$$

where j is a form factor to normalise the area to 1. There are two strong deviations from the distribution function in Fig. 5.6 (right). At the beginning of the main linac $\rho(E)$ is Gaussian and much wider, with a standard deviation σ of 0.02 (see Fig. 5.6 (left)). At the end of the main linac the distribution function is similar to the average one, but slightly narrower. These deviations will be dealt with in the Sec. 5.3.4.

5.3.3. Calculation of the basic amplitude model

5.3.3.1. Analytic expressions of the beam motion

Combining Eqs. (5.19), (5.24) and (5.26), assuming a unit excitation ($A_k = 1$) and making the substitution $u = \hat{\phi}(s)(1 - f\delta)$ gives the integral

$$Y(s > s_k) = r \int_{u_{down}}^{u_{up}} \sin(u)(ou^2 + pu + q)du \quad \text{with} \quad (5.27)$$

$$o = \frac{l}{\hat{\phi}(s)^2 f^2}, \qquad r = -\frac{A(s)}{\hat{\phi}(s) f A(s_k)},$$

$$p = -\frac{2l\hat{\phi}(s)}{\hat{\phi}(s)^2 f^2} + \frac{m}{\hat{\phi}(s) f}, \qquad q = \frac{l\hat{\phi}(s)^2}{\hat{\phi}(s)^2 f^2} - \frac{m\hat{\phi}(s)}{\hat{\phi}(s) f} + n,$$

$$u_{up} = \hat{\phi}(s)(1 - 0.0063f), \qquad u_{down} = \hat{\phi}(s)(1 + 0.0045f),$$

$$l = \frac{g}{j}, \qquad m = -2\frac{gh}{j},$$

$$\text{and} \qquad n = \frac{gh^2}{j} + \frac{i}{j}.$$

5.3. Amplitude model for the main linac of CLIC

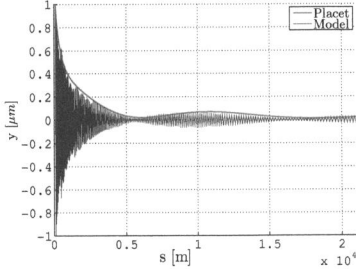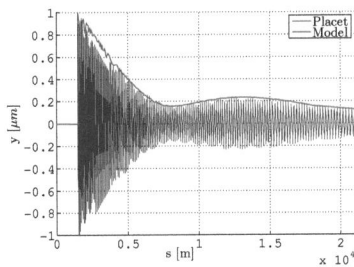

Figure 5.7.: Beam oscillations due to kicks in the main linac at $s_k = 100$ m (blue, left) and $s_k = 1500$ m (blue, right) and the according fits of the amplitude model (red). For the kick at $s_k = 100$ m, the filamentation is very strong and the particle oscillations become quickly completely incoherent. For the kick at $s_k = 1500$ m the filamentation effect is not as strong as for the kick at $s_k = 100$ m, but the amplitude of the beam oscillations is still dominated by the incoherence.

Integration by parts gives the final result

$$Y(s > s_k) = r((2ou_{up} + p)\sin(u_{up}) - (ou_{up}^2 + pu_{up} + q - 2o)\cos(u_{up}) \\ - (2ou_{down} + p)\sin(u_{down}) + (ou_{down}^2 + pu_{down} + q - 2o)\cos(u_{down})). \quad (5.28)$$

5.3.3.2. Amplitude extraction

Equation (5.28) contains amplitude and phase information. However, just the amplitude information is meaningful due to the made approximations. Since Eq. (5.28) could not be simplified further, the amplitude $Y_{amp}(s)$ was extracted with the help of a peak detection algorithm.

The algorithm first identifies all positive peaks by checking, if the actual point has a larger amplitude as both, the next point and the point before. For the detection of the negative peaks, the same criterion is applied to the inverted data. The absolute values of the detected peaks are smoothed by applying a Hamming low pass filter. This reduces the model error, since the oscillation amplitudes are mainly smooth along the main linac. The created amplitude data are sampled at the BPM locations. The described procedure is computationally very fast. An example of the produced oscillation envelopes is shown in Fig. 5.7 (right).

5.3.4. Extensions to the basic amplitude model

The basic oscillation amplitude model derived in Sec. 5.3.3 is not valid, if strong energy distribution variations occur in the main linac. While these effects can be neglected at the end of the main linac, they cause the large errors in the basic model predictions, if the beam is kicked at the beginning of the main linac. The reason for this modelling problem is that at the beginning of the main linac particles with the same energy and

5. System identification scheme for orbit response matrices

at the same position can have different phases. This was excluded in the basic model by the assumption, that the δ of a particle stays constant along the main linac. To still be able to produce the necessary amplitude information, a twofold approach is used. In case $s_k > 1000\,\text{m}$, the basic model is used. For kicks at $s_k \leq 1000\,\text{m}$, the following fit to simulation data from PLACET is employed.

A typical beam oscillation at the beginning of the main linac has a form depicted in Fig. 5.7 (blue curve, left). It has two minima z_1 and z_2 and three maxima m_0, m_1 and m_2. Simulations showed that these turning points can be approximated with

$$\begin{aligned}
s_{z1} &= 6000 + 2s_k, & s_{z2} &= 17500 + 2s_k, \\
y_{z1} &= 1.5 \times 10^{-4} s_k + 0.007, & y_{z2} &= 1.1 \times 10^{-4} s_k + 0.003, \\
s_{m0} &= s_k, & s_{m1} &= (s_{z1} + s_{z2})/2, \\
s_{m2} &= 21000, & y_{m0} &= 1, \\
y_{m1} &= 0.0019 s_k^{2/3} + 0.025, & y_{m2} &= 1.2 \times 10^{-4} s_k + 0.015.
\end{aligned} \qquad (5.29)$$

Based on these positions, a fit of the form

$$\hat{Y}(s > s_k) = \frac{\alpha(s) * \beta_1(s) + \beta_2(s)}{\gamma_1(s)} \qquad (5.30)$$

is calculated. The components of Eq. (5.30) will be explained shortly. The basic behaviour of the repeating minima and maxima is modelled by

$$\alpha(s) = \cos(\omega s + \varphi_0)^2 \quad \text{with} \qquad (5.31)$$
$$\omega = \pi/(s_{z2} - s_{z1}) \quad \text{and} \quad \varphi_0 = \pi/2 - \omega s_{z1}.$$

To lift the minima to their proper position, the linear function

$$\beta_2(s) = \frac{y_{z1} - y_{z2}}{s_{z1} - y_{z2}} s + \frac{s_{z1} y_{z2} - s_{z2} y_{z1}}{s_{z1} - y_{z2}} \qquad (5.32)$$

is added. To account for the decreasing amplitude of the maxima,

$$\beta_1(s) = \frac{1}{\hat{m}_1 s^4 + \hat{m}_2 s^2 + \hat{m}_3} \qquad (5.33)$$

is multiplied by $\alpha(s)$, where $\hat{m}_1$, $\hat{m}_2$ and $\hat{m}_3$ result from the solution of the following system of linear equations

$$\begin{bmatrix} s_k^4 & s_k^2 & 1 \\ s_{m1}^4 & s_{m1}^2 & 1 \\ s_{m2}^4 & s_{m2}^2 & 1 \end{bmatrix} \begin{bmatrix} \hat{m}_1 \\ \hat{m}_2 \\ \hat{m}_3 \end{bmatrix} = \begin{bmatrix} \frac{1}{1 - \beta_2(s_k)} \\ \frac{1}{y_{m1} - \beta_2(s_{m1})} \\ \frac{1}{y_{m2} - \beta_2(s_{m2})} \end{bmatrix}. \qquad (5.34)$$

In order to increase the slope between s_k and s_{z1}, the function $\gamma_1(s)$ is used, which is given by

$$\gamma_1(s) = \begin{cases} 1, & 0 \leq s \leq s_k, \\ \left(\dfrac{\hat{b}s}{\hat{c}s^2 + 1} + 1\right)\hat{i}s & s_k \leq s \leq s_e, \\ 1, & s_e \leq s \leq s_{\text{end}}, \end{cases} \quad \text{with} \qquad (5.35)$$

128

5.3. Amplitude model for the main linac of CLIC

$$\hat{b} = \frac{2y_f}{s_f}, \qquad \hat{c} = \frac{1}{s_f^2}, \qquad s_e = 6000 + s_k,$$

$$s_f = \frac{200}{\sqrt{10}}\sqrt{s_k} + 1000, \qquad y_f = \frac{2.8\sqrt{\hat{a}}}{\sqrt{s_k + \hat{a}}}, \qquad \hat{a} = \frac{400}{2.8^2 - 1}.$$

The linear function

$$\hat{l}(s) = \hat{e}(s - s_k) + 1 \quad \text{with} \quad \hat{e} = \frac{1}{s_e}\left(\frac{1}{\hat{f}} - 1\right) \quad \text{and} \quad \hat{f} = \frac{\hat{b}s_e}{\hat{c}s_e^2 + 1} + 1$$

ensures that at s_{end} the effect of the correction function dies out smoothly. The resulting $\hat{Y}(s)$ for a beam kick at e.g. $s_k = 200\,\text{m}$ can be seen in Fig. 5.7 (red line, left).

The effect of the strong change in the particle energy distribution for kicks at the beginning of the main linac does not die out exactly at 1000 m. Especially at the end of the main linac, residual errors can be seen. To reduce the model error further, the correction function

$$\gamma_2(s) = \begin{cases} 0, & 0 \leq s < 18000 \\ \hat{g}s + \hat{h}, & 1800 \leq s \leq s_{end} \end{cases} \quad \text{with} \tag{5.36}$$

$$\hat{g} = 1.33 \times 10^{-5} \quad \text{and} \quad \hat{h} = -0.24 \tag{5.37}$$

was added to the basic model in Eq. (5.28).

5.3.5. Amplitude model validation

To verify the accuracy of the described amplitude model, the produced data were compared to simulation results. For this reason, beam oscillations generated with the simulation code PLACET were processed by the same peak detection algorithm as used for the amplitude model. The resulting envelope $Y_{\text{sim}}(s_k, s)$ was subtracted from the according amplitude model data, which forms an error vector $e(s_k, s) = Y(s_k, s) - Y_{\text{sim}}(s_k, s)$. The relative quadratic error of all generated error vectors

$$\Delta = \frac{\sqrt{\sum_{k=1}^{N} e(s_k, s)^T e(s_k, s)}}{\sqrt{\sum_{k=1}^{N} Y_{\text{sim}}(s_k, s)^T Y_{\text{sim}}(s_k, s)}} \tag{5.38}$$

is used as an accuracy measure, where N is the number of positions at which the beam is kicked. Two evaluations were performed for different kick positions. In the first

	Δ_{model}	Δ_{ext}	Δ_{ML}
s_k every 21 m	3.4 %	8.6 %	3.6 %
s_k every QP	3.9 %	9.2 %	4.5 %

Table 5.1.: Accuracy evaluation of the amplitude model.

5. System identification scheme for orbit response matrices

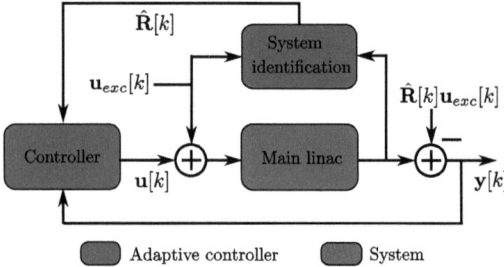

Figure 5.8.: Combination of the system identification and the L-FB. Additionally to the actuation $u[k]$ of the L-FB, the excitation $u_{exc}[k]$, created by the excitation unit of the system identification algorithm, is applied to the accelerator inputs. The L-FB should not react on the beam oscillations created by the system identification, since these oscillations will be changed from time step to time step. Therefore, the predicted beam oscillations due to the system identification $\hat{R}[k]u_{exc}[k]$ are removed from the measurements before the measurement vector $y[k]$ is fed back to the L-FB. The estimate $\hat{R}[k]$ of the orbit response matrix $R[k]$ can be used to update the controller coefficients of the L-FB. The overall system is hence an adaptive control system.

evaluation, the beam is kicked at every quadrupole in the beam line. This puts more emphasis on the beginning of the main linac, since the quadrupoles are closer together in this area. It is also the relevant value for the accuracy of an estimated response matrix. In the second evaluation, the beam is kicked every 21 m. For every evaluation, Δ is calculated for the overall model, but also separately for the fit at the beginning of the main linac ($s_k \leq 1000$) and the model for the main part of the main linac ($s_k > 1000$). The resulting values are presented in Tab. 5.1.

The accuracy of the overall model is around 4 %. Considering the necessary simplifications (e.g. constant particle distribution, continuous beta function), this result is very good. While the error of the main model is even lower (around 3.5 %), the fit at the beginning of the main linac is not that accurate (about 9 %). The reason for the larger error are the quickly changing envelope functions in this region, which are difficult to fit. However, since the region in which the fit is applied is short compared to the overall length of the main linac, these inaccuracies do not play a critical role for the accuracy of the full model.

5.4. Simulations results of the identification scheme with the L-FB

To use the presented system identification algorithm for the main linac of CLIC, the system has to be combined with the L-FB. For the following simulations, the weighted SVD controller presented in Sec. 3.2.2.2 is used as the L-FB for the main linac, which is a simplified version of the orbit controller for the main linac and the BDS. The combination of the system identification and the orbit controller is shown in Fig. 5.8.

5.4. Simulations results of the identification scheme with the L-FB

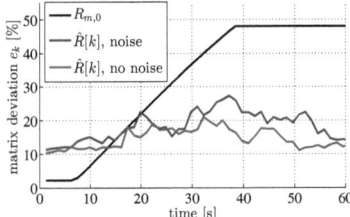 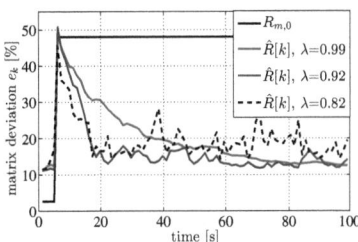

Figure 5.9.: Quadratic error of the initially measured orbit response matrix $R_{m,0}$ and different estimated matrices $\hat{R}[k]$ compared to the real matrix $R[k]$ of the main linac, subject to a change of the acceleration gradients. The acceleration gradients are altered in a coherent way by 0.15 MV/m and -0.3° along the main linac, which is a possible scenario given the specified gradient tolerances. (left) The gradient change occurs in a drift like manner from 6.20 s to 38.44 s and produces an error between the measured $R_{m,0}$ and the real $R[k]$ (black curve) of about 48 %. With the system identification algorithm ($\lambda = 0.92$) the change can be learned and the final error is about 13 %. The blue curve corresponds to the error of the estimated $\hat{R}[k]$ with BPM noise (50 nm) and ground motion (model B, quadrupole stabilisation V1), while for the red curve no disturbances have been applied. (right) The same gradient change as for the left plot has been applied in a step like manner at 6.20 s. Learning curves for different forgetting factors λ are shown, where disturbances and noise has been applied.

The estimated response matrix estimate $\hat{R}[k]$ can not only be used for system diagnosis, but also for an update of the parameters of the L-FB (recalculation of the SVD). This corresponds to an adaptive control scheme, which improves the performance of the L-FB in case of system changes over time. More precisely, the shown adaptive control scheme belongs to the group of *self tuning regulators* (STR), which is characterised by the fact that the system identification and the control algorithm are not combined to one algorithm, but separately realised.

The adaptive controller was implemented in PLACET and Octave and tested via simulations. The results are presented in Fig. 5.9. While a drift like change of the acceleration gradients (left) results in a large error for an initially measured response matrix $R_{m,0}$, the identified matrix can adapt to the system change. Note that also the measured response matrix has an initial error (in this setup 2.7 %) due to the present measurement noise. The identified matrix possesses a steady state error of about 13 % (in the current setup), which is composed of two components. The first component accounts for 10.8 % error and is due to the limited accuracy of the amplitude model and the simplified assumption of a smooth phase advance along the linac (modelling errors). The second component is due to the influence of measurement noise and ground motion. This component depends on the choice of the forgetting factor λ. A larger value (closer to 1) corresponds to stronger averaging and thus to a small influence of disturbances. On the other hand, a high λ also results in a slow learning speed. This trade-off between learning speed and steady state error is illustrated in Fig. 5.9 (right) and Tab. 5.5.

5. System identification scheme for orbit response matrices

α	0.99	0.97	0.95	0.92	0.90	0.87	0.85	0.82	0.79	0.75
Δe_d [%]	1.0	1.0	2.1	2.1	3.0	4.2	4.7	7.0	6.4	7.7
$T_{90\%}$ [s]	50.8	34.7	26.0	19.8	17.4	14.9	13.6	12.4	11.2	9.9

Table 5.2.: Trade-off between the steady state error Δe_d due to disturbances and learning speed $T_{90\%}$ for different forgetting factors α. The value $T_{90\%}$ is the time needed to learn 90 % of a step like change of the orbit response matrix. The values of Δe_d are not strictly monotonically increasing in the table due to statistical fluctuations.

5.5. Conclusions

In this chapter, it was shown how methods from system identification theory can be used to identify time-changing orbit response matrices of accelerators. In general, the full orbit response matrix can be identified with high precision, if enough time is available. In the presented work, such a basic identification algorithm (RLS) has been extended to speed up the identification by a factor of 33. In return, the modified scheme can not completely identify all changes in the orbit response matrix anymore, but is optimised to identify phase advance deviations, which are assumed to be the most significant error sources. The modified system identification algorithm combines a performance optimised RLS algorithm with a phase reconstruction and merging algorithm and an amplitude model for beam oscillations in the main linac of CLIC.

In particular, the amplitude model is an independent tool itself. It includes the effects of the beta function and filamentation on the beam oscillations. The model assumes constant relative energy deviation along the main linac for every particle. At the beginning of the main linac, where this assumption is not true, the model was exchanged with a fit to simulated data. An overall accuracy of about 4 %, and a calculation time of below 0.15 s on a standard PC with low memory consumption have been achieved. This shows that the developed model is accurate, fast and efficient. Improvements in the calculation speed are assumed to be possible, since a simple non-optimised Matlab script was used. While the fit in the beginning of the main linac is specific for CLIC, the model for the main part can be adapted to other accelerators, which use a beam with an approx. constant relative energy deviation.

Simulations showed that the complete system identification algorithm is able to detect steps and drifts of accelerator parameters, in spite of measurement noise and ground motion disturbances. The trade-off between steady state error and identification speed has been evaluated as a function of the forgetting factor λ. The choice of $\lambda = 0.92$ turned out to be a good compromise for the parameters of the main linac. With this forgetting factor, 90 % of a step-like change in the orbit response matrix can be learned within about 20 s with a steady state error of about 13 %. It has also been shown in this chapter how the system identification scheme can be combined with the operation of the L-FB.

The presented scheme is flexible and can be adapted easily to the parameters of other accelerators. For most machines it should be possible to reduce the steady state error of the identification compared to the main linac of CLIC, since the main linac of CLIC is a particularly inconvenient system to model and excite.

6. Conclusions

In this thesis, we present essential contributions to the feasibility issue of the luminosity preservation of CLIC with respect to ground motion effects. The achievements of this work can be grouped in three parts: controller design, integrated simulations and system identification, which will be detailed in the following paragraphs. As a result of these efforts, the feasibility issue of luminosity preservation due to ground motion has been resolved even with relatively pessimistic assumptions for the ground motion model.

Controller design: Standard orbit feedbacks are not able to cope with the demanding requirements necessary to control the beam orbit in the main linac and the beam delivery system (BDS) of CLIC. The problem arises since the according obit feedback, named linac feedback (L-FB), has to suppress the ground motion effects very efficiently, while at the same time being robust against measurement noise of the beam position monitors (BPMs). Therefore, a novel design method has been developed in Sec. 3.2. It is based on a decoupling controller, which is a well-known strategy for orbit feedbacks (SVD controller). The innovation of the design is the novel method to determine the controller for the individual decoupled channels. A two step, semi-automatic approach is taken. First, the user defines a basic transfer function, which allows him to incorporate important expert knowledge about the accelerator system. Then, an automatic algorithm calculates an open parameter of the given user transfer function such that the luminosity loss, or some other target function, is minimised. The algorithm uses therefor models for the assumed ground motion excitation and measurement noise introduced in Sec. 2.1 and 2.2.

The tuned controller preserves the luminosity even better than a similar, hand optimised orbit feedback controller, which was adapted over several weeks to achieve its maximum performance. The design time has been drastically reduced. Additionally, due to the improved robustness against measurement noise the tight specification for the BPM resolution could be relaxed from 10 nm to 50 nm in the BDS and possibly even to 100 nm in the main linac. The presented design method is an interesting option for future orbit feedback systems including storage rings.

Also simple but effective designs for the interaction point feedback (IP-FB) have been presented in Sec. 3.4. The according controller coefficients have been optimised with the help of derived models (Sec. 2.3) and the achieved performance is close to the theoretical optimum. Also cost reduction options for the quadrupole stabilisation system have been investigated (Sec. 3.3). No strong limitations for these systems have been found.

Integrated simulations: A simulation framework has been set up to design, test and evaluate all ground motion mitigation methods of CLIC (Sec. 2.4). The framework combines and extends existing codes: beam tracking with PLACET, luminosity calculation with GUINEA-PIG, controller and algorithm implementation in Octave and a realistic

6. Conclusions

ground motion generator. Such computationally expensive, full-scale simulations are necessary, since the interplay of the accelerator, ground motion and mitigation methods cannot be described with sufficient accuracy by simple analytical formulas.

Using the basic stabilisation system (V1) and a pessimistic assumption for the ground motion model (model B10), simulations presened in Sec. 4.1 showed that the luminosity loss of CLIC can be limited to a level of about 12 %. This result is sufficient but at the limit of the allowable tolerance. Further investigations showed that the simulation results can be improved, if the frequency response of the quadrupole stabilisation system is changed in some frequency ranges. The according guidelines for the modification of the frequency response were used by the stabilisation group to design an improved system named V2. With this stabilisation system the luminosity loss is as small as 2 %, including the measurement noise.

The overall system was tested in Sec. 4.2 with respect to several machine imperfections. The only problem that has been observed is a sensitivity due to beam energy variations. These variations create large transversal offsets in the dispersive area of the BDS and are measured in the according BPMs. The measurements are coupled back by the L-FB and create a significant luminosity loss. The problem was resolved by adding a small modification, named dispersion filter, to the L-FB. The dispersion filter removes the dispersive orbit in a very efficient way from the BPM measurements. The remaining luminosity loss due to beam energy variations is negligible. Additionally, the dispersion filter can be used to measure the beam energy with potentially very high accuracy.

System identification scheme: The third part of this thesis is concerned with the problem of obtaining high quality system knowledge of the accelerator, which is in this case the orbit response matrix. A system identification scheme is presented in Chap. 5 that is capable of identifying changes of this matrix over time, without interrupting physics data taking (on-line). A standard algorithm (RLS) has been modified to speed-up the identification process by a factor of 33. This was achieved by focusing on the main source of changes of the orbit response matrix, while neglecting less important disturbances. As a consequence a steady state error of about 13 % has to be accepted. However, if the speed-up is not necessary, the identification scheme can be modified to identify the full orbit response matrix with high precision. The overall scheme also includes a phase reconstruction algorithm, which calculates from measured BPM readings the effective phase advance of the accelerator, and a derived amplitude model of beam oscillations in the main linac. The amplitude model includes the lattice dependence of the oscillations as well as the energy dependent effect of filamentation.

Simulations have shown that the system identification scheme works, also in combination with the L-FB. The system can reduce the down time of accelerators, since it is not necessary anymore to interrupt the physics operation to measure the response matrix. The identified orbit response matrix can be used for several applications. By updating the controller parameters of the orbit controller with the identified matrices (adaptive controller), the feedback performance can be improved. The matrices are also an important input for beam-based alignment methods, system diagnosis and error detection tools. Since the identification scheme is very flexible it can be easily adapted to other machines.

A. Beam physics background

In this appendix, some basic concepts and terms from the field of beam physics are covered. This brief overview is addressed to people with a different background and should help to make this thesis better accessible for them. For much more detailed introductions into the field of beam physics, please refer to the texts Wiedemann [135] and Wille [136]. Be aware that in this thesis also concepts are used, which are not covered here. In such cases we will refer to the relevant literature.

A.1. Luminosity

The *luminosity*, symbolised by $\mathcal{L}$, is together with the particle energy the most important accelerator parameter. It is a measure for the quality of the beam collisions in a particle accelerator and is proportional to the rate of particle production, which is given by

$$\frac{\mathrm{d}N}{\mathrm{d}t} = \sigma(E)\mathcal{L}, \tag{A.1}$$

where N is the total number of produced particles due to a certain reaction and $\sigma(E)$ is the *cross section* for this reaction. The cross section of a reaction describes the probability for this reaction to occur. It is dependent on the type, state and energy of the colliding particles and has the unit of an area.

The luminosity describes the efficiency of the particle beam collisions. In a linear collider and for Gaussian distributed beams, the luminosity can shown to be (Napoly [76])

$$\mathcal{L} = \frac{f_{rep} n_b N_b^2}{4\pi \sigma_x \sigma_y}, \tag{A.2}$$

where f_{rep} is the beam train repetition rate, n_b is the number of bunches per beam train, N_b is the number of particles in a bunch and σ_x and σ_y are the beam sizes (standard deviation of the Gaussian distribution). CLIC aims for a total luminosity of 5.9×10^{34} cm^{-2}s^{-1}. For physics research, the peak luminosity $\mathcal{L}_{1\%}$ that is generated only from particles within 1 % of the nominal collision energy is more relevant. The target value for $\mathcal{L}_{1\%}$ at CLIC is 2×10^{34} cm^{-2}s^{-1} and whenever we use the term luminosity in this thesis we will refer to the peak luminosity.

Imperfections, as e.g. ground motion, can decrease the nominal luminosity $\mathcal{L}_0$ by a value $\Delta\mathcal{L}$. We will in the following state expressions for the luminosity decrease due to beam size growth and beam-beam offset, which are used in Chap. 2 for the creation of models.

In case the beam size deviates from its nominal values $\sigma_{x,0}$ and $\sigma_{y,0}$ by $\Delta\sigma_x$ and $\Delta\sigma_y$, the relative luminosity loss is given by

$$\frac{\Delta\mathcal{L}}{\mathcal{L}_0} = 1 - \frac{\sigma_{x,0}\sigma_{y,0}}{(\sigma_{x,0}+\Delta\sigma_x)(\sigma_{y,0}+\Delta\sigma_y)} \approx \frac{\Delta\sigma_x}{\sigma_{x,0}} + \frac{\Delta\sigma_y}{\sigma_{y,0}} + \mathcal{O}(2). \tag{A.3}$$

A. Beam physics background

In Eq. (A.3) the last approximation is due to a Taylor expansion, where terms of order two and higher have been neglected. In case the two beams collide with offsets δ_x and δ_y, the luminosity loss can be calculated as

$$\frac{\Delta \mathcal{L}}{\mathcal{L}_0} = 1 - e^{-\frac{1}{4}\left[\left(\frac{\delta_x}{\sigma_{x,0}}\right)^2 + \left(\frac{\delta_y}{\sigma_{y,0}}\right)^2\right]} \approx \left(\frac{\delta_x}{2\sigma_{x,0}}\right)^2 + \left(\frac{\delta_y}{2\sigma_{y,0}}\right)^2 + \mathcal{O}(4), \qquad (A.4)$$

where again a Taylor series expansion has been used to obtain the last approximation. The Eqs. (A.2), (A.3) and (A.4) neglect the mutual interaction of the two beam at the collision. To also take into account these beam-beam interactions, non-Gaussian beam distributions and the beam energy spread due to Beamstrahlung, numerical simulation tools as GUINEA-PIG (see Schulte [108]) have to be used. GUINEA-PIG is used as an element of the developed integrated simulation framework in Sec. 2.4.

A.2. Emittance and beam size

When a beam moves along the accelerator with length s, it is usually described with the help of the statistical properties of its individual particles. Each individual particle (here indexed with i) is defined by its horizontal, vertical and longitudinal positions x_i, y_i, z_i; by the vertical and horizontal "angle" in propagation direction $x'_i = \mathrm{d}x_i/\mathrm{d}s$ and $y' = \mathrm{d}y_i/\mathrm{d}s$ and its energy E. These particle coordinates can be collected in the vector $\boldsymbol{x}_i = [x_i \, x'_i \, y_i \, y'_i \, z_i \, \delta_i]^T$, where $\delta_i = (E_i - E_0)/E_0$ is the relative energy deviation of the particle and $E_0 = \mathbb{E}\{E_i\}$ is the average beam energy. Since the beam is assume to be Gaussian distributed, it can be described by the first and second order moments of its particle coordinate distributions as

$$X = \mathbb{E}\{\boldsymbol{x}_i\} \quad \text{and} \quad \Sigma = \mathbb{E}\left\{\boldsymbol{x}_i \boldsymbol{x}_i^T\right\}, \qquad (A.5)$$

where Σ is the so called *6D beam matrix*. For sake of simplicity, we restrict ourselves for a moment only on the vertical dimension (the other dimensions are analogue), which leads to the 2D beam matrix

$$\Sigma_y = \begin{bmatrix} \mathbb{E}\{y_i y_i\} & \mathbb{E}\{y_i y'_i\} \\ \mathbb{E}\{y'_i y_i\} & \mathbb{E}\{y'_i y'_i\} \end{bmatrix} = \epsilon_y \begin{bmatrix} \beta_y & -\alpha_y \\ -\alpha_y & \gamma_y \end{bmatrix}. \qquad (A.6)$$

When describing the beam transport along an accelerator it is convenient to separate the properties of the beam and the magnet system of the accelerator called lattice. This separation is given in Eq. (A.6) by the *emittance* ϵ_y, which is a beam property, and the *twiss parameters* α_y, β_y and γ_y, which are given by the lattice. The emittance is an invariant (without acceleration) of the beam and describes how easily the beam can be focused to small beam sizes. From Eq. (A.6) we can see that the beam size can be calculated from the emittance and the twiss parameters as

$$\sigma_y = \sqrt{\epsilon_y \beta_y}. \qquad (A.7)$$

Note that the beam size is not only dependent on the emittance but also on the lattice design and thus not an appropriate quantity to describe the beam independent from the magent system.

Without acceleration the beam emittance is preserved in an accelerator, neglecting dissipative effects like synchrotron radiation. If a beam is accelerated on the other hand, it shrinks due to a relativistic effect called adiabatic damping. For the lattice design it is often of advantage to work with the *normalised emittance* ϵ_N, which stays constant even when the beam is accelerated. The normalised emittance is defined as

$$\epsilon_N = \gamma_\epsilon \epsilon \quad \text{with} \quad \gamma_\epsilon = \frac{E_0(s)}{m_0 c^2},$$

where m_0 is the rest mass of the accelerated particle type. In contrast to the normalised emittance ϵ_N, ϵ is sometimes called geometric emittance. Another important term is the *projected emittance* ϵ_p. The beam size can only be calculated by Eq. (A.7), if the 6D beam matrix is decoupled in horizontal, vertical and longitudinal direction. If this is not the case, also coupling from other directions has to be taken into account, which leads to the projected emittance. This projected emittance is what actually can be measured and what determines the usable beam quality. In contrast to the normalised emittance, the projected, normalised emittance it is not constant along the accelerator. When we use the term emittance in this thesis, we refer implicitly to the projected, normalised emittance. This concept can be further generalised, by not considering only one but several beam bunch. This leads to the so called *multi-bunch emittance*, which includes also the offset of consecutive bunches.

Finally we want to state how the beam size changes, if it is composed out of two individual contribution. We assume therefor that additionally to the nominal beam size σ_0 we have a small additional beam size contribution σ_c, due to coupling or dispersion (see next section for a definition). In this case the overall beam size σ is given by

$$\sigma = \sqrt{\sigma_0^2 + \sigma_c^2}.$$

The beam size change due to σ_c can be approximated as

$$\Delta\sigma = \sigma - \sigma_0 \approx \frac{\sigma_c^2}{\sigma_0} + \mathcal{O}(3), \tag{A.8}$$

where a Taylor expansion has been used. The expression Eq. (A.8) is used in Sec. 2.3.2 for the development of a model of the beam size growth due to an offset of the final doublet magnets.

A.3. Energy spread and dispersion

In this section we discuss the effects arising from the fact that the particles of a beam do not all have the same energy but are Gaussian distributed with a standard deviation δ usually called energy spread. The different particle energies cause the particles to be differently influenced by the magnets of the accelerator lattice. Particles with higher energies will be less deflected by the magnetic fields compared to particles with lower energies. As a result, the natural oscillations along the beam line of the particles with higher energies will have longer wave lengths and higher amplitudes.

If a beam passed through a dipole field, particles with different energies will have the different deflection angles. As a result, a dependence of the transversal position

A. Beam physics background

of a particle on the energy can be observed. The change of transversal position along the beam line due to an energy deviation can be calculated (here only for the vertical direction) as

$$y_D(\delta, s) = d_y(s)\delta \quad \text{with} \quad d_y(s) = d_y = \frac{\mathbb{E}\{y\delta\}}{\mathbb{E}\{\delta^2\}}, \tag{A.9}$$

where $d_y(s)$ is called *dispersion* at the position s and δ is the relative energy deviation of the particle. Note that not only dipole magnets can create dipole field, but also quadrupoles in case the beam enters these quadrupoles with an offset. As an example, the offset of the final focus quadrupoles can create dispersion that increases the beam size at the IP strongly (see Sec. 2.3.2).

Beside dispersion, the natural energy spread of the beam can have another beam quality decreasing effect. If a beam is kicked as a whole, e.g. by a misaligned quadrupole magnet, it oscillates along the beam line. Since particles with higher energy oscillate slower than particle with lower energy, a phase difference will build up and the beam motion gets less and less coherent. As a result, the oscillation of the beam centre is damped, but at the same time the beam emittance grows. This effect is called *filamentation* and a model for it is derived in Sec. 5.3 for the main linac of CLIC.

B. Integrated simulation framework

In Sec. 2.4, an integrated simulation framework was described for the numerical analysis of the complex effects of ground motion and the ground motion mitigation systems in the main linac and BDS of CLIC. In this appendix we give additional information for the usage of this framework.

B.1. Download and installation

Before the simulation framework can be set up, the freely available tools PLACET, GUINEA-PIG and Octave have to be installed. PLACET and GUINEA-PIG can be downloaded via the *Concurrent Version Service* (CVS) with the following commands.

```
export CVSROOT=:ext:anonymous@isscvs.cern.ch:/local/reps/placet
export CVS_RSH=ssh
cvs checkout placet-development
cvs checkout gp
```

Here the useage of the *bash shell* is assumed. If another shell is used, the environment variables CVSROOT and CVS_RSH have to be adapted appropriately. For the installation of PLACET we refer to the documentation placet-development/doc/placet.pdf. Note that not only the executable placet-octave, but also the created program grid has to be copied to a directory in the search path of the shell, e.g. /usr/bin/. To install GUINEA-PIG, the FFTW libraries, where FFTW stands for "Fastest Fourier Transform in the West", have to be downloaded and installed. We recomment the version 2.1.5 of the FFTW libraries. With these files, GUINEA-PIG can be compiled by simply typing make in its root directory. Octave can easily be found in the Internet, but we would like to mention that due to the current implementation of PLACET, the version of Octave has to be between 2.1.17 and 3.0.5 and for the creation of the make file with ./configure the option --without-hdf5 should be used. After these tools are installed the framework itself can be checked out by

```
export CVSROOT=:ext:anonymous@isscvs.cern.ch:/local/reps/placet
export CVS_RSH=ssh
cvs checkout clic-integrated-simulations
```

B.2. Usage and interface

The simulation framework can be started with the following commands.

```
cd clic-integrated-simulations/linac-bds/dynamic/
placet-octave run_linac_bds_integrated.tcl user_settings.tcl params
```

B. Integrated simulation framework

In the last line, PLACET executes the central script run_linac_integrated.tcl from which all other function are called. All used functions and necessary data files are collected in the folder dynamic/scripts/. The functions are implemented in the two languages Tcl/Tk and Octave (file endings .tcl and .m), but only basic knowledge of these scripting languages is necessary to understand and extent the code.

The first step in the execution of the central script is to load the settings for the simulation. The default settings are located in the file settings.tcl, which is loaded first. After that the user-defined settings file user_settings.tcl is evaluated and parameters from settings.tcl are overwritten with the parameters of user_settings.tcl. It is further possible to pass an unlimited number of parameters param, which have to be applied to simulation parameters in user_settings.tcl by manually written code. A user settings file user_settings.tcl is written in the language Tcl/Tk and has for example the following form.

```
# handle command line parameter

if {$argc >= 2} {
    set groundmotion(seed_nr) [lindex $argv 1]
}
if {$argc >= 3} {
    puts "ERROR: too many arguments"
    exit
}

# Overwrite default settings:

set ground_motion_x 1
set ground_motion_y 1
set groundmotion(model) "D"

set nr_time_steps [expr 50*60]
set delta_T 0.02
dir_name "home_dir"

# Feedbacks

set use_controller_x 1
set use_controller_y 1
set use_ip_feedback_x 1
set use_ip_feedback_x 1

set bpm_noise 1
```

In the first few lines the input parameters argv are processed. The number of input parameters argc is always one higher than the number of parameters, since the user settings file is also counted as one parameter. In this settings file it is assumed that no or one parameter is passed to the script, in order to initialise the ground motion generator seed groundmotion(seed_nr). If two or more parameters are passed, the program execution is stopped. After this input parameter handling, the settings for the simulation are defined. In this case, ground motion of model D is used (groundmotion(model)) in the horizontal (ground_motion_x) and vertical (ground_motion_y) direction. A total of 50*60 beam bunches (nr_time_steps) are simulated with a separation of delta_T=0.02 s (repetition rate of CLIC). This corresponds to a real-time operation of 60 s. As working directory, home_dir is used. For the simulations, the L-FB and the IP-FB in the horizontal (use_controller_x and use_ip_feedback_x) and vertical (use_controller_y and use_ip_feedback_y) directions are switched on. BPM noise is considered in the simulations (bpm_noise).

The execution of the simulation scripts will produce typically three output files with the names meas_station_1/2/3_maschine_1.dat. The index 1 corresponds to a virtual measurement station after the main linac, at which certain results are recorded. The first column of the according file is the time, will for the meaning of the other columns we refer the user to the implementation in dynamic/scripts/postprocess_time_step.tcl. The indices 2 and 3 correspond to similar virtual measurement stations before the final doublet and at the IP. Additional outputs can be generated by switching on flags in the settings file.

B.3. Parameters

In Sec. B.2, we mentioned already some of the many simulation parameters of the integrated simulation framework. All parameters and their default values are listed in the file settings.tcl and we will in the following explain the most important once. Terms as x/y in the parameter names will refer to two parameters, one for x and one for y.

General simulation parameters

- delta_T: Time between two simulation steps (usually the beam repetition time)

- nr_machines: Not only one, but multiple simulations with different seeds can be performed sequentially (default is 1)

- nr_time_steps: Number of time steps to be simulated

- dir_name: Name of the working directory, where simulations outputs are stored

- use_main_linac: Determines if the main linac should be included in the simulations

- use_bds: Determines if the BDS should be included in the simulations

- use_beam_beam: Determines if the beam collisions should be simulated (GUINEA-PIG)

- debug: Produces additional outputs for debugging

- response_matrix_calc: Special mode in which not a usual simulation is performed, but the orbit response matrix is calculated

B. Integrated simulation framework

Ground motion parameters

- ground_motion_x/y: Turns on ground motion in the horizontal and/or vertical direction

- groundmotion(type): Type of the ground motion: 0 ... non, 1 ... ATL, 2 ... standard models

- groundmotion(model): Determines the ground motion model if standard models are used: A, B, C, D; where here D is used for B10

- groundmotion(filtertype): Use of the stabilisation system or the pre-isolator

- groundmotion(filtertype_x/y): File where the frequency response of the stabilisation system or the pre-isolator is stored

- groundmotion(preisolator): Type of used pre-isolator; see settings.tcl for an listing of the options

L-FB and IP-FB parameters

- use_controller_x/y: Use L-FB

- controller_type_spatial_x/y: Type of spatial L-FB filter: $1 \ldots f_i = 1$, 2 and 3 ... hand-optimised f_i, 4 ... automatically optimised f_i

- gain_file_name_x/y: File containing the automatically optimised gains

- controller_type_frequency_x/y: Type of time-dependent L-FB filter: 1 ... integrator, 2 ... integrator and low pass, 3 ... full time-dependent filter

- use_disp_suppression_x/y: Use dispersion suppression

- use_ip_feedback_x/y: Use IP-FB

- use_ip_feedback_linear: Use simple proportional IP-FB

- use_ip_feedback_pid: Use PID IP-FB

- use_ip_feedback_annecy: Use adaptive IP-FB designed from SYMME and LAPP

Imperfection parameters

- bpm_noise: Use of BPM noise in the main linac and BDS

- bpm_resolution_ml/bds: Resolution of the BPMs in the main linac and BDS in [μm]

- bpm_ip_noise: Use of BPM noise in the post-collision BPM for the IP-FB

- bpm_ip_res: Resolution of the BPM in the post-collision line in [μm]

- **stabilization_noise_x/y**: Use of noise from the quadrupole stabilisation (more specific parameters in `settings.tcl`)
- **use_rf_jitter**: Use of acceleration gradient jitter (more specific parameters in `settings.tcl`)
- **use_qp_jitter**: Use of quadrupole strength jitter (more specific parameters in `settings.tcl`)

Beam, lattice and beam-beam parameters

- **params**: Array of parameters, which determine the initial beam properties as emittance and energy spread
- **e_initial_linac**: Initial beam energy at the entrance of the main linac
- **e_initial_bds**: Initial beam energy at the entrance of the BDS, in case the main linac is not used in the simulations
- **n_slice** and **n_part**: Number of slices and particles per slice of the macro-particle beam used for the tracking in the main linac
- **n_total**: Number of particles used for the tracking in the BDS
- **gp_param**: Array of parameters relevant for the execution of GUINEA-PIG

Post-processing and simulation specific parameters

Post-processing parameters determine, which simulation results should be stored in addition to the standard output files. Simulation specific parameters are only relevant for special applications in which the user has added non-standard functionality to the framework.

C. Control engineering background

In this appendix, a brief introduction in the control engineering methods use in this thesis will be given. The presented material is intended to make this thesis accessible to people not familiar with the field of control engineering. For more detailed information about the topic, please refer to the literature that will be mentioned.

C.1. $\mathcal{Z}$-transform

The $\mathcal{Z}$-transform is a mathematical formalism that transforms discrete-time signals and time-discrete linear systems into an equivalent frequency domain representation. It is therefore very similar to the Laplace-transform, which is used for continuous-time signals and systems. The $\mathcal{Z}$-transform is an important method for the design of discrete-time control systems.

The $\mathcal{Z}$-transform $X(z)$ of a discrete-time signal $x[k]$, where $z \in \mathbb{C}$ and $k \in \mathbb{Z}^+$ is the time index, is given by

$$X(z) = \mathcal{Z}\left\{x[k]\right\} = \sum_{k=0}^{\infty} x[k] z^{-k}, \tag{C.1}$$

which is the definition of the $\mathcal{Z}$-transform. In case $x[k]$ is the impulse response of a linear system, $X(z)$ is also called the discrete-time transfer function. To distinguish signals and systems, we will use the terms $X(z)$ and $Y(z)$ for signals and $S(z)$ for systems. In the following, some properties of the $\mathcal{Z}$-transform are stated, which are of importance for this thesis. For more detailed information, please refer to the standard texts Oppenheim, Schafer and Buck [78] and Franklin, Powell and Workman [42].

It is often necessary to extract the frequency response from a transfer function $S(z)$. This can be achieved by evaluating $S(z)$ on the unit circle of the z-plane, i.e. $z = e^{j\omega T_d}$, where T_d is the sampling time of the discrete-time system. Since $e^{j\omega T_d}$ is a periodic function with respect to frequency, also the frequency response of every discrete-time system is periodic. For comparison, the Laplace-transformed $S(s)$ of a system has to be evaluated on the imaginary axis, i.e. $s = j\omega$, to obtain the frequency response. Another important property is that a time-discrete linear system is stable, if and only if, all poles of its transfer function $S(z)$ lie inside the unit circle of the z-plane. This is in contrast to the Laplace-transform where, a system is stable, if and only if, all poles of $S(s)$ lie left of the imaginary axis.

Also very useful is the ability to create time domain realisations from the discrete-time transfer function of a system. This is necessary, since the design of filters (system) is often performed in the frequency domain, while for its implementation a realisation in the time domain is needed. To carry out the necessary inverse $\mathcal{Z}$-transform $\mathcal{Z}^{-1}\left\{.\right\}$, only a few basic rules have to be known. A time shift in the time domain $y[k] = x[k+i]$ has

C. Control engineering background

a frequency domain representation of $Y(z) = z^i X(z)$. This representation of the time shift operator as a multiplication with z^i is used frequently in this thesis. By further mentioning that the $\mathcal{Z}$-transform and its inverse are linear operations and that filter transfer functions $S(z)$ are ratios of polynomial functions in our application, the time domain representation of $S(z)$ can be written as

$$\mathcal{Z}^{-1}\left\{Y(z) = S(z)X(z)\right\}$$
$$\Rightarrow \mathcal{Z}^{-1}\left\{Y(z) = \frac{a_m z^m + a_{m-1} z^{m-1} + \cdots + a_1 z + a_0}{z^n + b_{n-1} z^{n-1} + \cdots + b_1 z + b_0} X(z)\right\} \quad \text{with} \quad m \leq n$$
$$\Rightarrow \mathcal{Z}^{-1}\left\{Y(z) + \cdots + \frac{b_1}{z^{n-1}} Y(z) + \frac{b_0}{z^n} Y(z) = \frac{a_m}{z^{n-m}} X(z) + \cdots + \frac{a_0}{z^n} X(z)\right\}$$
$$\Rightarrow y[k] = a_m x[k-(n-m)] + \cdots + a_0 x[k-n] - b_{n-1} y[k-1] - \cdots - b_0 y[k-n]. \quad \text{(C.2)}$$

C.2. Controller design using the frequency domain

In this section, basic knowledge about controller design in the frequency domain will be covered. Design methods in the frequency domain do not determine the controller with the help of the time domain system representation, which are differential or difference equations. Instead, the system to be controlled is transformed in a frequency representation where the controller design is carried out. For continuous systems this transformation is given by the Laplace-transform, while for discrete-time systems the $\mathcal{Z}$-transform (see Sec. C.1) is used. Since this thesis is concerned with discrete-time systems, the explanations will be restricted to discrete-time transfer functions. Additionally, only single-input, single-output systems (SISO) will be considered. In the following, we cover the topics of the standard control loop, the loop shaping technique (including Nyquist's stability theorem) and performance limitations of feedback control. Aspects that are of importance for this thesis are emphasised. Only a brief overview about these topics can be given here and we refer the interested reader for more information to the texts Franklin, Powell and Workman [42] and Gausch, Hofer and Schlacher [46].

C.2.1. Standard control loop

Figure C.1 shows the standard control loop in its frequency domain representation and defines the involved signals and systems. Only one dynamic element $C(z)$ is employed in this configuration to control the system $H(z)$. Note that in the standard literature instead of $H(z)$ often the expressions $G(z)$ or $P(z)$ are used, which are already occupied by different quantities in this thesis. Many control configurations can be posed in this simple form, as e.g. the L-FB system for the main linac and BDS of CLIC. In the case of the L-FB, $H(z)$ corresponds to the accelerator system, $D(z)$ to ground motion, $N(z)$ to BPM noise and $R(z)$ to the reference orbit. The output signal $Y(z)$ can be calculated

C.2. Controller design using the frequency domain

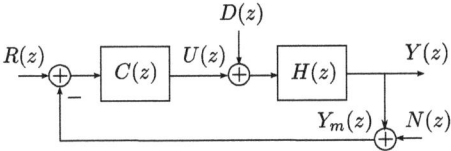

Figure C.1.: Block diagram of the standard control loop in frequency domain representation. The output signal $Y(z)$ of the system $H(z)$ has to be controlled by the controller $C(z)$. Typically $Y(z)$ has to follow the reference signal $R(z)$ as good as possible and should be influenced by the disturbance signal $D(z)$ and the measurement noise $N(z)$ as little as possible. To fulfil these specifications, the controller has only the measured output signal $Y_m(z)$ available, which includes the noise $N(z)$.

from the input signals of the control loop as

$$Y(z) = H(z)\left[D(z) + C(z)\left(R(z) - N(z) - Y(z)\right)\right] \Rightarrow$$
$$Y(z) = H(z)S(z)D(z) + T(z)R(z) - T(z)N(z) \quad \text{with} \tag{C.3}$$
$$S(z) = \frac{1}{1+H(z)C(z)} \quad \text{and} \quad T(z) = \frac{H(z)C(z)}{1+H(z)C(z)}, \tag{C.4}$$

where $S(z)$ and $T(z)$ are called the sensitivity and the complementary sensitivity function. The sensitivity function $S(z)$ for the standard control loop should not be confused with other sensitivity functions $G(k)$ of e.g. the luminosity loss or the beam-beam offset as introduced in Chap. 2.

Many important properties of $\mathcal{Z}$-transformed signals and systems become apparent, if they are evaluated at $z = e^{j\omega T_d}$. This leads to the definition of the frequency response $H(z = e^{j\omega T_d}) = H(e^{j\omega T_d})$, where T_d is the sampling time of the discrete-time system. A physical interpretation of the frequency response is that a harmonic signal $A\sin[\omega k T_d + \phi_0]$, where k symbolises the time index of a discrete signal in this case, is changed by passing through the system $H(e^{j\omega T_d})$ to

$$A\sin[\omega k T_d + \phi_0] \xrightarrow{H(e^{j\omega T_d})} \left|H(e^{j\omega T_d})\right| A\sin\left[\omega k T_d + \phi_0 + \arg\left(H(e^{j\omega T_d})\right)\right].$$

Note that the frequency response for a discrete-time system is a periodic function with respect to ω. All harmonic functions with angular frequency $\omega_n = \omega_0 + 2\pi n/T_d$, $n \in \mathbb{Z}$, are treated by the system described by $H(e^{j\omega T_d})$ the same way.

C.2.2. Loop shaping

Loop shaping is a design method that aims to chose the controller $C(e^{j\omega T_d})$ such that the complementary sensitivity function $T(e^{j\omega T_d})$ and the sensitivity function $S(e^{j\omega T_d})$ obtain a desirable shape. Before the discussion of the design of such a controller $C(e^{j\omega T_d})$, it has to be define which shapes for $T(e^{j\omega T_d})$ and $S(e^{j\omega T_d})$ are desirable. For the measurement noise suppression, it is ideal if $T(e^{j\omega T_d})$ is equal to 0. For the reference signal tracking, on the other hand, a $T(e^{j\omega T_d})$ of 1 results in the perfect behaviour, since all reference

C. Control engineering background

signals at the input are exactly reproduced at the output. In general, a trade-off has to be found between these two conflicting demands. In this thesis however, the reference signal tracking is not of high importance, which simplifies the choice of $T(e^{j\omega T_d})$.

The ideal choice of the sensitivity function $S(e^{j\omega T_d})$, which determines the disturbance suppression, is equal to 0. However, also in this case a compromise has to be found, since the disturbance suppression is not independent of the noise behaviour and the here less important reference signal tracking. This can be seen from the relation $S(e^{j\omega T_d}) + T(e^{j\omega T_d}) = 1$, which can be easily verified from the definition of $S(e^{j\omega T_d})$ and $T(e^{j\omega T_d})$ in Eq. (C.4). We face the situation in which only one controller $C(e^{j\omega T_d})$ has to fulfil multiple objectives. In the following, we will get to know additional restrictions that limit the choice of the controller further.

After the desirable shapes of $S(e^{j\omega T_d})$ and $T(e^{j\omega T_d})$ have been determined, the design of a controller that achieves these open loop shapes can be discussed. To understand the shaping of $T(e^{j\omega T_d})$ and $S(e^{j\omega T_d})$, it is helpful to introduce the open loop frequency response $L(e^{j\omega T_d}) = C(e^{j\omega T_d})H(e^{j\omega T_d})$, which should not be confused with the frequency response of the low pass $L(z)$ in Sec. 3.2.3.2. With the help of $L(e^{j\omega T_d})$, the functions $S(e^{j\omega T_d})$ and $T(e^{j\omega T_d})$ can be approximated as

$$S(e^{j\omega T_d}) = \frac{1}{1+L(e^{j\omega T_d})} \approx \begin{cases} \frac{1}{L(e^{j\omega T_d})} & \text{if } |L(e^{j\omega T_d})| \gg 1 \\ 1 & \text{if } |L(e^{j\omega T_d})| \ll 1 \end{cases} \tag{C.5}$$

$$T(e^{j\omega T_d}) = \frac{L(e^{j\omega T_d})}{1+L(e^{j\omega T_d})} \approx \begin{cases} 1 & \text{if } |L(e^{j\omega T_d})| \gg 1 \\ 0 & \text{if } |L(e^{j\omega T_d})| \ll 1. \end{cases} \tag{C.6}$$

Since we are mainly interested in this thesis in the disturbance and noise suppression, we focus on these aspects in the following and neglect the reference signal tracking.

From the approximation in Eq. (C.5), it is clear that for a good disturbance rejection $|L(e^{j\omega T_d})| \gg 1$ is needed. This can be achieved by using a controller with high gain. However, there are several constraints that limit the maximum controller gain. The first limitations can be deducted from the approximation Eq. (C.6). If a large $|L(e^{j\omega T_d})|$ is used also the noise is transferred without demagnification to the output of the system. The second limitation comes from the fact that a high controller gain will result in large controller outputs (actuator set point), which can excees the limits of the actuators. Therefore, the controller gain has to be reduced to appropriate values.

The last limitation is due to stability reasons, expressed by *Nyquist's stability theorem*. We will here only state a simplified version of this theorem, which is sufficient to analyse the stability behaviour of the L-FB in this thesis. We restrict ourselves to discrete-time, linear, SISO systems of the form

$$L(z) = k\frac{1}{(z-1)^p}\frac{M(z)}{N(z)} \quad \text{with}$$
$$M(z=1) = N(z=1) = 1, \quad k>0 \quad \text{and} \quad p \in \{0,1,2\}, \tag{C.7}$$

where $M(z)$ and $N(z)$ are polynomials in z and all zeros of $N(z)$ have to be inside the unit circle. The magnitude $|L(e^{j\omega T_d})|$ is assumed to be 1 only for one so called *cross over angular frequency* ω_c in the interval $[0, \pi/T_d]$. Futhermore, $|L(e^{j\omega T_d})|$ is assumed to be larger 1 for all $\omega < \omega_c$ and smaller 1 for all $\omega > \omega_c$ in the interval $[0, \pi/T_d]$. To

C.2. Controller design using the frequency domain

exclude the possibility of full encirclements of the unit circle, the phase of $L(e^{j\omega T_d})$ is assumed to be in the interval $[-\pi, 3\pi/2]$ for all $\omega < \omega_c$ in the interval $[0, \pi/T_d]$. If we define the *phase margin* as $\phi_m = \pi + \arg\left(L(e^{j\omega_c T_d})\right)$, Nyquist's stability theorem for this system type states that the closed loop system is stable if and only if $\phi_m > 0$.

The implications of Nyquist's stability theorem, limit the gain of a controller by the following reasoning. The use of a high gain controller will in general lift the curve $|L(e^{j\omega T_d})|$, which increases the cross over frequency ω_c and therefore reduces the phase margin ϕ_m. As an additional effect, controller elements that are used to increase the open loop gain produce often additional phase shift (e.g. integrators), which reduces the phase margin even more. Hence, high gain feedback controller can bring the closed loop system closer to its stability limits and create eventually instabilities. Since, especially for high frequencies, a certain mismatch between the system model and the real system has to be expected, the phase margin of the design should be larger than approx. $\pi/6$, which limits the maximum gain.

Concluding we can state that a good shaped open loop frequency response, should have high gain for low frequencies, where only small model errors have to be expected and the phase shift of the open loop is small. At high frequencies the gain has to be reduced to keep the system stable. The cross over frequency may additionally be reduced, if the design amplifies noise too strongly and/or produces too high actuator set points. The loop shaping technique is used in Sec. 3.2.3.2 (see Fig. 3.18) to design the time-dependent filter for the L-FB.

C.2.3. Bode's sensitivity integral

It is important for the designer of feedback systems to understand the basic limitations of feedback control. Such knowledge prevents the designer form demanding an unfeasible performance of the controller. For the disturbance rejection, such a performance limitation is given by *Bode's sensitivity integral*. This theorem is a well-known result for the continuous case (see Stein [127]), but can also be extended to discrete-time systems (see Mohtadi [74]). For a standard control loop with a stable transfer function, Bode's sensitivity integral for discrete-time systems is given by

$$\frac{1}{\pi} \int_0^\pi \ln \left| S(e^{j\phi}) \right| d\phi = \sum_{i=1}^m \ln |\bar{p}_i|, \qquad (C.8)$$

where $\bar{p}_i$ are the unstable poles of the open loop system and m is the total number of these points. Note that a pole excess of two, as it is necessary for continuous systems, is not required in the time-discrete case.

Bode's sensitivity integral states that the area below the logarithmically measured magnitude of the sensitivity function $S(e^{j\omega T_d})$ is a constant. This constant is dependent on the number and position of the unstable poles of the open loop system. For an open loop transfer function with all poles inside or on the unit circle, as it is the case in this thesis, the area is zero. This implies that if a disturbance is demagnified due to feedback control over some frequency range, it will be amplified over another frequency range. Disturbance suppression cannot be achieved for over the full frequency range with feedback control. This does however not imply that feedback control is not effective for

C. Control engineering background

disturbance suppression. In practical applications, $S(e^{j\omega T_d})$ is chosen such that the disturbances with the highest impact on the output are demagnified. The inevitable disturbance amplification is positioned at less significant frequency ranges.

C.3. Kalman-filtering

We assume in the following that the reader is familiar with the state space representation of linear dynamic systems. Comprehensive introductions to the topic are given in Oppenheim, Willsky and Hamid [79] and Dourdoumas and Horn [32]. Consider a discrete-time, linear, dynamic system in state space representation

$$x[k+1] = Ax[k] + Bu[k] + w[k] \tag{C.9}$$
$$y[k] = Cx[k] + n[k], \tag{C.10}$$

where $u[k]$ and $y[k]$ are the inputs and outputs of the system at time step k. Beside the deterministic signal $u[k]$, there are two stochastic processes present: the state disturbance $w[k]$ changes the system states $x[k]$ in a stochastic manner and the measurement noise $n[k]$ disturbs the output measurements $y[k]$. Both, $w[k]$ and $n[k]$, are assumed to be white, Gaussian distributed, stochastic processes with zero mean value and covariance matrices W and N.

A problem of great practical interest is the estimation of the hidden states $x[k]$ from the known inputs $u[k]$ and the measurements $y[k]$. Such an estimate $\hat{x}[k]$ can be used as an input for a state controller or for diagnosis tools. The estimation algorithm that solves this problem is a copy of the real dynamic system in Eqs. (C.9) and (C.10) and is called *observer*. The mismatch between the estimated output of the observer $\hat{y}[k]$ and the real output $y[k]$ is used to change the states of the observer $\hat{x}[k]$, such that the variance of the error $e[k] = \hat{x}[k] - x[k]$ is minimised. To be able to build such an observer the system to be observed has to have a property called observability. For criteria to verify if a system is observable, we refer to Skogestad and Havre [123].

The observer design problem is hindered by the fact that the measurements $y[k]$ are not only dependent on the states $x[k]$, but also on the measurement noise $n[k]$. If the estimate of the last time step $\hat{x}[k-1]$ was already close to $x[k-1]$, a too strong use of the noisy $y[k]$ could increase the estimation error again. A well balanced trade-off between the quality of the already gained estimate $\hat{x}[k]$, the amount of noise $n[k]$ and the size of the state change due to $w[k]$ has to be found. The observer that produces the optimal estimate (in a quadratic sense) for this problem is called *Kalman-filter*. This Kalman-filter consists of the following 5 equations.

$$x^*[k] = A\hat{x}[k-1] + Bu[k-1] \tag{C.11}$$
$$P^*[k] = AP[k-1]A^T + W \tag{C.12}$$
$$K[k] = P^*[k]C^T \left(CP^*[k]C^T + N\right)^{-1} \tag{C.13}$$
$$\hat{x}[k] = x^*[k] + K[k]\left(y[k] - Cx^*[k]\right) \tag{C.14}$$
$$P[k] = \left(I - K[k]C\right)P^*[k] \tag{C.15}$$

For a prove that verifies that the Eqs. (C.11)-(C.15) produce such an optimal estimate, we refer to Grewal and Andrews [48] and Kalman [61]. In case an estimate for the next time step $\hat{x}[k+1]$ should be calculate, the Kalman-filter can be modified to a *Kalman-predictor* that calulates $\hat{x}[k+1]$ with the simply extrapolation $\hat{x}[k+1] = A\hat{x}[k]$.

C.4. Singular value decomposition

The singular value decomposition (SVD) splits up any matrix $A \in \mathbb{R}^{m \times n}$ into three sub-matrices as

$$A = U_A \Sigma_A V_A^T \quad \text{with} \quad U_A \in \mathbb{R}^{m \times m}, \quad \Sigma_A \in \mathbb{R}^{m \times n} \quad \text{and} \quad V_A \in \mathbb{R}^{n \times n}. \quad (C.16)$$

Both, U_A and V_A are orthonormal matrices. Orthonormal matrices have the important property that $U_A^T U_A = I$, where I is the identity matrix. The matrix Σ_A is a diagonal matrix, with the singular values $s[i]$ of A as the diagonal elements, ordered from the largest $s[1]$ to the smallest $s[\min(n, m)]$.

The SVD is a very useful tool, due its ability to give important insights in the linear transformation represented by A. If a vector c is multiplied by A, the multiplication can be split up into

$$Ac = U_A \Sigma_A V_A^T c = \left(\sum_i s[i] u[i] v[i]^T \right) c, \quad (C.17)$$

where $u[i]$ and $v[i]$ are the ith column of U_A and V_A. From Eq. (C.17), it is evident that the linear transformation A can be separated into a sum of sub-transformations consisting of the triplet $v[i]$, $u[i]$ and $s[i]$, where i is the sub-transformation index. The magnitude of $s[i]$ determines the importance of the sub-transformation for the overall transformation. The vector $v[i]$ determines for which direction of the vector c this sub-transformation (scaling by $s[i]$ and rotation to the direction $v[i]$) applies. This ability of the SVD do distinguish important from less important input directions can be used in many applications as e.g. data compression algorithms, conditioning number improvement of matrices and design of decoupling controller. Another useful property of the SVD is that the pseudo inverse of A can be efficiently calculated by

$$A^\dagger = V_A \Sigma_A^{-1} U_A^T. \quad (C.18)$$

More detailed information about the SVD can be found in Golub and van Loan [47].

C.5. RLS algorithm with exponential forgetting

The *recursive least squares algorithm* (RLS) is a tool to calculate unknown system parameters θ from known input and output signals $\varphi[k]$ and $y[k]$, where k is the time index. A memoryless system structure of the form

$$y[k] = \theta_1 \varphi_1[k] + \theta_2 \varphi_2[k] + \cdots + \theta_n \varphi_n[k] + n[k] = \boldsymbol{\theta}^T \boldsymbol{\varphi}[k] + n[k] \quad \text{with} \quad (C.19)$$
$$\boldsymbol{\theta} = [\theta_1, \theta_2, \ldots, \theta_n]^T \quad \text{and} \quad \boldsymbol{\varphi} = [\varphi_1[k], \varphi_2[k], \ldots, \varphi_n[k]]^T$$

is assumed, where $n[k]$ represents measurement noise that is modelled as a white, zero mean, Gaussian stochastic process. The system Eq. (C.19) is a multiple input, single

C. Control engineering background

output system (MISO system) with no internal dynamics. Generalisations of the below described RLS algorithm for systems with internal back coupling are available but not used in this thesis.

A commonly used approach to find parameter estimates $\hat{\boldsymbol{\theta}}$ for the real parameter $\boldsymbol{\theta}$ is to minimise the cost function

$$J(\hat{\boldsymbol{\theta}}, k) = \frac{1}{2} \sum_{i=1}^{k} \left(y[i] - \hat{\boldsymbol{\theta}}^T \boldsymbol{\varphi}[i] \right)^2. \tag{C.20}$$

The term $e[i] \equiv y[i] - \hat{\boldsymbol{\theta}}^T \boldsymbol{\varphi}[i]$ corresponds to the error between the real and estimated measurements. By minimising $J(\hat{\boldsymbol{\theta}}, k)$, also $e[i]$ is minimised in a quadratic sense, which leads to the well known *least squares solution*

$$\hat{\boldsymbol{\theta}}[k] = \left(\boldsymbol{\Phi}[k]^T \boldsymbol{\Phi}[k] \right)^{-1} \boldsymbol{\Phi}[k]^T \boldsymbol{Y}[k] \quad \text{with} \tag{C.21}$$

$$\boldsymbol{\Phi}[k] = \begin{bmatrix} \boldsymbol{\varphi}[1]^T \\ \vdots \\ \boldsymbol{\varphi}[k]^T \end{bmatrix} \quad \text{and} \quad \boldsymbol{Y}[k] = \begin{bmatrix} y[1] \\ \vdots \\ y[k] \end{bmatrix}. \tag{C.22}$$

The term $\left(\boldsymbol{\Phi}[k]^T \boldsymbol{\Phi}[k] \right)^{-1} \boldsymbol{\Phi}[k]^T$ is called the pseudo-inverse of $\boldsymbol{\Phi}[k]$. Note that in Eq. C.21, the full pseudo-inverse has to be recalculated, if new measurements are available. Since this recalculation is computational expensive (especially if many measurements are available) Eq. (C.21) is not well suited for on-line applications. Therefore, it is convenient to convert Eq. (C.21) into a recursive algorithm that delivers the least squares solution without recalculating the full pseudo-inverse in every time step. This recursive algorithm is called *recursive least squares algorithm* (RLS) and will be presented with a small additional modification shortly.

Till now, the system parameters $\boldsymbol{\theta}$ have been assumed to be constant. If the system parameters are slowly changing over time $\boldsymbol{\theta}[k]$, it is necessary to weight new measurements stronger than older ones to be able to "forget" the older system parameters over time. This can be accomplished by changing the cost function $J(\hat{\boldsymbol{\theta}}, k)$ in Eq. (C.20) to

$$J_\lambda(\hat{\boldsymbol{\theta}}, k) = \frac{1}{2} \sum_{i=1}^{k} \lambda^{k-i} \left(y[i] - \hat{\boldsymbol{\theta}}^T \boldsymbol{\varphi}[i] \right)^2, \tag{C.23}$$

where the term λ^{k-i} is an exponential function that is intended to weight more recent measurements stronger than older ones. The factor λ is called learning factor and has to be chosen between 0 and 1. Minimising the cost function $J_\lambda(\hat{\boldsymbol{\theta}}, k)$ and stating the solution in a recursive way results in the *recursive least squares algorithm with exponential forgetting* that has the form

$$\hat{\boldsymbol{\theta}}[k] = \hat{\boldsymbol{\theta}}[k-1] + \boldsymbol{K}[k](y[k] - \boldsymbol{\varphi}[k]^T \hat{\boldsymbol{\theta}}[k-1]) \tag{C.24}$$

$$\boldsymbol{K}[k] = \boldsymbol{P}[k-1]\boldsymbol{\varphi}[k](\lambda \boldsymbol{I} + \boldsymbol{\varphi}[k]^T \boldsymbol{P}[k-1]\boldsymbol{\varphi}[k])^{-1} \tag{C.25}$$

$$\boldsymbol{P}[k] = (\boldsymbol{I} - \boldsymbol{K}[k]\boldsymbol{\varphi}[k]^T)\boldsymbol{P}[k-1]/\lambda. \tag{C.26}$$

For proofs and intuitive interpretations of Eqs. (C.24), (C.25) and (C.26), we refer to the texts Åström and Wittenmark [95] and Ljung and Gunnardsson [70].

Bibliography

[1] International Linear Collider Technical Review Committee: Second Report. Technical Report SLAC Report-606, SLAC, 2003.

[2] In *Proceedings of the CAS - CERN Accelerator School: Basic Course on General Accelerator Physics*, Frascati (Italy), 2008.

[3] E. Adli. *A Study of the Beam Physics in the CLIC Drive Beam Decelerator*. PhD thesis, University of Oslo, 2009.

[4] K. Aki and P. G. Richards. *Quantitative Seismology (Second edition)*. University Science Books, 2009. ISBN: 1-89-138963-7.

[5] K. Artoos et al. Ground Vibration and Coherence Length Measurements for the CLIC Nano-Stabilization Studies. In *Proceedings of the 2009 Particle Accelerator Conference (PAC99)*, 2009.

[6] B. A. Baklakov et al. Investigation of seismic vibrations and relative displacements of linear collider VLEPP elements. In *Proceedings of the 1991 Particle Accelerator Conference (PAC91)*, 1991.

[7] V. Balakin, S. Novpkhatsky, and Smirnov V. VLEPP: transverse beam dynamics. In *Proceedings of the 12th International Conference on High Energy Accel.*, 1983.

[8] G. Balik et al. Interaction point feedback design and integrated simulations to stabilize the CLIC final focus. In *Proceedings of the 2nd International Particle Accelerator Conference (IPAC11)*, 2011.

[9] G. Balik et al. Integrated simulation of ground motion mitigation techniques for the future compact linear collider (CLIC). *Nuclear Instruments and Methods in Physics Research Section A*, 700:163 – 170, 2013.

[10] D. S. Barr. An adaptive feedback controller for transverse angle and position jitter correction in linear particle beam accelerators. *Given at 1992 Accelerator Instrumentation Workshop*, Berkeley, CA, 27-30 Oct 1992.

[11] M. Battaglia, A. de Roeck, J. Ellis, and D. Schulte. *Physics at the CLIC Multi-TeV Linear Collider : report of the CLIC Physics Working Group*. CERN, Geneva, 2004.

[12] B. Bolzon. *Etude des vibrations et de la stabilisation à l'échelle sous-nanométrique des doublets finaux d'un collisionneur linéaire*. PhD thesis, Université de Savoie, 2007.

Bibliography

[13] E. Bozoki and A. Friedman. Neural networks and orbit control in accelerators. In *Proceedings of the 1994 European Particle Accelerator Conference (EPAC94)*, pages 1589–1591, 1994.

[14] D. Brandt et al. A new closed-orbit correction procedure for the CERN SPS and LEP. *Nuclear Instruments and Methods in Physics Research A*, 293(1-2):305 – 307, 1990.

[15] H. Braun et al. CLIC 2008 Parameters. Technical report, The CLIC Study Team, 2008. http://cdsweb.cern.ch/record/1132079/files/CERN-OPEN-2008-021.pdf.

[16] K. L. Brown. A First- and Second-Order Matrix Theory for the Design of Beam Transport Systems and Charged Particle Spectrometers. Technical Report SLAC Report 75, SLAC, 1982.

[17] D. Bulfone. Overview of fast beam position feedback systems. In *Proceedings of the 2008 European Particle Accelerator Conference (EPAC08)*, 2008.

[18] P.N. Burrows et al. Performance of the FONT3 fast analogue intra-train beam-based feedback system at ATF. In *Proceedings of the 2006 European Particle Accelerator Conference (EPAC06)*, 2006.

[19] P.N. Burrows et al. The FONT4 ILC intra-train beam-based digital feedback system prototype. In *Proceedings of the 2007 Particle Accelerator Conference (PAC07)*, 2007.

[20] B. Caron, G. Balik, and L. Brunetti. Adaptive vibration control of the beam of the future linear collider. *Control and Engineering Practice*, 2011. to be published.

[21] A. Chao and M. Tigner. *Handbook of accelerator physics and engineering*. World Scientific, 1999. ISBN: 9-81-023500-3.

[22] A. W. Chao. *Physics of Collective Beam Instabilities in High Energy Accelerators*. John Wiley & Sons, Inc., 1993. ISBN: 0-471-55184-8.

[23] Y. Chung. Beam Position Feedback System for the Advanced Photon Source. In *Proceedings of the Orbit Correction and Analysis Workshop*, pages 155–162, 1994.

[24] C. Collette et al. Active quadrupole stabilization for future linear particle colliders. *Nuclear Instrumentation and Methods in Physics Research A*, to be published:0–0, 2010.

[25] C. Collette et al. Seismic response of linear accelerators. *Physical reviews special topics - accelerators and beams*, 2010.

[26] C. Collette et al. Nano-motion control of heavy quadrupoles for future particle colliders: An experimental validation. *Nuclear Instruments and Methods in Physics Research A*, (643):95 – 101, 2011.

[27] C. Collette et al. Review of sensors for low frequency seismic vibration measurement. Technical report, CERN, 2011. CERN-ATS-Note-2011-001.

Bibliography

[28] R. Corsini and J. P. Delahaye. The CLIC Multi-Drive Beam Scheme. Technical Report CLIC-Note-331, CERN, 1998.

[29] B. Dalena et al. Beam delivery system tuning and luminosity monitoring in the Compact Linear Collider. *Physical Review Special Topics - Accelerators and Beams*, 15(5), 2012.

[30] M. Dehler. Real-time control of beam parameters. In *Proceedings of the CERN Accelerator School on Digital Signal Processing*, 2007.

[31] R. C. Dorf and R. H. Bishop. *Modern Control Systems*. Prentice Hall, 2008. ISBN: 0-132-27028-5.

[32] N. Dourdoumas and M. Horn. *Regelungstechnik*. Pearson Studium, 2003. ISBN: 3-8273-7059-0.

[33] J. C. Doyle, B. A. Francis, and A. Tannenbaum. *Feedback Control Theory*. Macmillan Pub. Co., 1992. ISBN: 0-023-30011-6.

[34] J. W. Eaton. Octave. Technical report. https://www.gnu.org/software/octave.

[35] P. Eliasson. Dynamic imperfections and optimized feedback design in the compact linear collider main linac. *Phys. Rev. Spec. Top. Accel. Beams*, 11:51003, 2008.

[36] P. Eliasson. *Emittance preservation and luminosity tuning in future linear colliders*. PhD thesis, Uppsala Universitet, 2008.

[37] J. Ellis. Particle physics at future colliders. In *Proceedings of the 26th Advanced ICFA Workshop on Nanometre-Size Colliding Beams Lausane*, 2002.

[38] J. Ellis and I. Wilson. New physics with the Compact Linear Collider. *Nature*, 409:431–435, 2001.

[39] G. E. Fischer. Ground motion and its effects in accelerator design. *SLAC-PUB-3319*, 1985.

[40] O. Föllinger. *Einführung in die Methoden und ihre Anwendung*. Hüthig Buch Verlag Heidelberg, 1994. ISBN: 3-7785-2915-3.

[41] G. Franklin, J. D. Powell, and A. Emami-naeini. *Feedback Control Of Dynamic Systems*. Prentice Hall, 2009. ISBN: 0-135-00150-1.

[42] G. Franklin, J. D. Powell, and M. L. Workman. *Digital Control of Dynamic Systems*. Prentice Hall, 1997. ISBN: 0-201-82054-4.

[43] A. Gaddi. Dynamic analysis of the final focusing magnets pre-isolator and support system. Technical Report LCD-Note-2010-11, CERN, 2010.

[44] A. Gamp. Servo control of RF cavities under beam loading. In *Proceedings of the CERN Accelerator School on RF Engineering for Particle Accelerators*, 1992.

Bibliography

[45] M. Gasior et al. Sub-nm beam motion analysis using a standard BPM with high resolution electronics. In *Proceedings of the 2010 Beam Instrumentation Workshop (BIW10)*, 2010.

[46] F. Gausch, A. Hofer, and K. Schlacher. *Digitale Regelkreise*. Oldenbourg Verlag, 1993. ISBN: 3-486-22734-3.

[47] G. H. Golub and Ch. F. van Loan. *Matrix Computations*. The Johns Hopkins University Press, 1996. ISBN: 0-801-85414-8.

[48] M. S. Grewal and A. P. Andrews. *Kalman Filtering: Theory and Practice Using MATLAB*. Wiley-IEEE Press, 2008. ISBN: 0-4701-7366-1.

[49] A. Grudiev and W. Wuensch. Design of an X-Band Accelerating Structure for the CLIC Main Linac. Technical Report CLIC-Note-773, CERN, 2008.

[50] M. Günther. *Zeitdiskrete Steuerungssysteme*. VEB Verlag Technik, 1988. ISBN: 3-341-00528-5.

[51] L. Hendrickson. Algorithms, Optimization and Simulation Results for Pulse-to-pulse feedback in SLC, NLC/JLC, CLIC and TESLA. In *the 33th Advanced Beam Dynamics Workshop (NANOBEAM02)*, 2002.

[52] L. Hendrickson et al. Fast feedback for linear colliders. In *Proceedings of the 1995 Particle Accelerator Conference (PAC95)*, pages 2389–2393, 1995.

[53] L. Hendrickson et al. Feedback Systems for Linear Colliders. In *Proceedings of the 1999 Particle Accelerator Conference (PAC99)*, pages 338–342, 1999.

[54] R.O. Hettel. Beam steering at the Stanford Synchrotron Radiation Labratory. *IEEE Transactions on Nuclear Science*, 30(2228-2230):305 – 307, 1983.

[55] T. Himel. Use of digital control theory state space formalism for feedback at SLAC. In *Proceedings of the 1991 Particle Accelerator Conference (PAC91)*, pages 1451–1453, 1991.

[56] T. Himel. Adaptive Cascaded Beam-Based Feedback at the SLAC. In *Proceedings of the 1993 Particle Accelerator Conference (PAC93)*, pages 2106–2108, 1993.

[57] T. Himel. Feedback: Theory and accelerator applications. *Annual Review of Nuclear and Particle Science*, 47:157–192, 1997.

[58] A. Hofmann. Landau damping. In *Proceedings of the CAS - CERN Accelerator School: Second Advanced Accelerator Physics Course*, Berlin, 1989. ISBN: 92-9083-004-2.

[59] S. Janssens et al. System control for the CLIC main beam quadrupole stabilization and nano-positioning. In *Proceedings of the 2nd International Particle Accelerator Conference (IPAC11)*, 2011.

[60] V. M. Juralev et al. Investigations of power and spatial correlation characteristics of seismic vibrations in the CERN LEP tunnel for linear collider studies. Technical Report CLIC-Note 217, European Organization for Nuclear Research (CERN), 1993.

[61] R. E. Kalman. A New Approach to Linear Filtering and Prediction Problems. *Transaction of the ASME, Journal of Basic Engineering*, 82:35 – 45, 1960.

[62] W. B. Klein, R. T. Westervelt, and G. F. Luger. Developing a general purpose intelligent control system for particle accelerators. *Journal of Intelligent and Fuzzy Systems*, 7/1:1–12, 1999.

[63] K. Kubo. Estimation of orbit change and emittance growth due to random misalignment in long linacs. *Phys. Rev. Spec. Top. Accel. Beams*, 4:14401, 2011.

[64] A. Kuzmin. Ground vibration measurements and Experiment parts motion measurement at CMS. Technical Report EDMS Nr :1027459, CERN, 2009.

[65] H. Kwakernaak and R. Sivan. *Linear Optimal Control Systems*. Wiley-Interscience, 1972. ISBN: 0-471-51110-2.

[66] F. Lackner et al. Technical Proposal: Laser Alignment Multipoint Based Design Approach (LAMBDA). Technical Report EDMS 1066954, CERN, 2010.

[67] A. Latina et al. Feedback studies. Technical report, EUROTeV, 2007. EUROTeV Report 2007 065.

[68] A. Latina and P. Raimondi. A novel alignment procedure for the final focus of future linear colliders. In *Proceedings of Linear Accelerator Conference (LINAC2010)*, 2010.

[69] N. Leros and D. Schulte. Dynamic effects in the main linac of CLIC. In *Proceedings of the 2001 Particle Accelerator Conference (PAC01)*, 2001.

[70] L. Ljung and S. Gunnardsson. Adaptation and Tracking in System Identification - A Survey. *Automatica*, 26(1):7–21, 1990.

[71] M. Lonza. Multi-bunch feedback systems. In *Proceedings of the CERN Accelerator School on Digital Signal Processing*, 2007.

[72] E. Meier et al. Artificial Intelligence Systems for Electron Beam Parameters Optimization at the Australian Synchrotron LINAC. In *Proceedings of the 1st International Particle Accelerator Conference (IPAC10)*, 2010.

[73] M. G. Minty and F. Zimmermann. *Measurement and Control of Charged Particle Beams*. Springer. ISBN: 9-783-54044187-8.

[74] C. Mohtadi. Bodes integral theorem for discrete time systems. *IEE Proc.*, 137(2):57 – 66, 1990.

Bibliography

[75] C. Montag. Active stabilization of mechanical quadrupole vibrations for linear colliders. *Nuclear Instruments and Methods in Physics Research A*, 378(3):369 – 375, 1996.

[76] O. Napoly. Beam Delivery System and Beam-Beam Effects. In *Proceedings of the ILC School 2009*, Beijing (China), 2009. www.linearcollider.org/school/2009/.

[77] A. Neumaier. Solving Ill-conditioned and Singular Linear Systems: A Tutorial on Regularization. *SIAM Review*, 4:636–666, 1998.

[78] A. V. Oppenheimer, R. W. Schafer, and J. R. Buck. *Discrete-time signal processing*. Prentice Hall, second edition, 1999. ISBN: 0-13-754920-2.

[79] A. V. Oppenheimer, A. S. Willsky, and S. Hamid. *Signals and Systems*. Prentice Hall, second edition, 1996. ISBN: 0-13-814757-4.

[80] A. Papoulis and S. U. Pillai. *Probability, random variables, and stochastic processes*. McGraw-Hill, 2002. ISBN: 0-07-112256-7.

[81] V. Parkhomchuk and V. Shiltsev. Fractal Model of Ground. In *Proceedings of the Int. Workshop on Linear Colliders*, 1992.

[82] M. Pezzetti. *Standard and Experimental Approach for Advanced Controls in Cryogenics*. PhD thesis, Université de Picardie de Jules Verne, 2010.

[83] J. Pfingstner et al. Adaptive Scheme for the CLIC Orbit Feedback. In *Proceedings of the 1st International Particle Accelerator Conference (IPAC10)*, 2010.

[84] J. Pfingstner et al. Amplitude model for beam oscillations in the main linac of CLIC. Technical Report CERN-OPEN-2011-010. CLIC-Note-860, CERN, 2010.

[85] J. Pfingstner et al. An interleaved, model-supported system identification scheme for the particle accelerator CLIC. In *Proceedings of the 49th IEEE Conference on Decision and Control (CDC10), to be published*, 2010.

[86] J. Pfingstner et al. Lockerung von Sensortoleranzen mittels regelungstechnischer Methoden für den Teilchenbeschleuniger CLIC. In *17. Steirisches Seminar über Regelungstechnik und Prozessautomatisierung*, 2011.

[87] J. Pfingstner et al. SVD-based filter design for the orbit feedback of CLIC. In *Proceedings of the 2nd International Particle Accelerator Conference (IPAC11)*, 2011.

[88] J. Pfingstner et al. Recent improvements of the orbit controller and ground motion mitigation techniques for CLIC. In *Proceedings of the 3rd International Particle Accelerator Conference (IPAC12)*, 2012.

[89] J. Pfingstner, J. Snuverink, and D. Schulte. Ground motion optimised orbit feedback design for the future linear collider. *Nuclear Instruments and Methods in Physics Research A*, 703:163 – 170, 2013.

Bibliography

[90] N. Phinney et al. International Linear Collider Reference Design Report, Volume 3: Accelerator. Technical report, 2007.

[91] M. Pieck. Artificial intelligence research in particle accelerator control systems for beam line tuning. In *Proceedings of LINAC08*, pages 314–316, 2008.

[92] M. A. Pinsky. *Introduction to Fourier Analysis and Wavelets*. American Mathematical Society, 2009. ISBN: 0-821-84797-X.

[93] A. Preumont. *Random Vibration and Spectral Analysis*. Springer Netherlands, 1994. ISBN: 0-79-233036-6.

[94] A. Preumont and K. Seto. *Vibration Control of active Structures an Introduction*. John Wiley & Sons Inc, 2008. ISBN: 0-470-03393-2.

[95] K. J. Åström and B. Wittenmark. *Adaptive Control*. Dover Publications, Inc., 2008. ISBN: 0-486-46278-1.

[96] T. O. Raubenheimer. *The generation and acceleration of low emittance flat beams for future linear colliders*. PhD thesis, Stanford University, 1991.

[97] T. O. Raubenheimer. Estimates of emittance dilution and stability in high-energy linear accelerators. *Phys. Rev. ST Accel. Beams*, 3(12):121002, 2000.

[98] T. O. Raubenheimer et al. Zeroth-order Design Report for the Next Linear Collider. Technical Report SLAC Report-474, SLAC, 1996.

[99] T. O. Raubenheimer and R. D. Ruth. A dispersion-free trajectory correction technique for linear colliders. *Nuclear Instruments and Methods in Physics Research Section A*, 302(2):191–208, 1991.

[100] T. O. Raubenheimer and P. Tenenbaum. Brief Review of Linear Collider Beam-Based Alignment for Linacs. Technical report, Stanford Linear Accelerator Center, 2004. SLAC-TN-03-071, LCC-0129.

[101] S. Redaelli. *Stabilization of Nanometer-Size Particle Beam in the Final Focus System of the Compact LInear Collider (CLIC)*. PhD thesis, Institut de Physique des Hautes Energies, Université de Lausanne, 2003.

[102] Y. Renier, P. Bambade, and A. Sery. Tuning of a 2D ground motion generator for ATF2. Technical Report LAL/RT 08-18, CARE/ELAN document-2008-005, ATF-08-10, LAL, 2008.

[103] J. Resta-López, P.N. Burrows, and G. Christian. Luminosity performance studies of the compact linear collider with intra-train feedback system at the interaction point. *Journal of Instrumentation*, 5(9), 2010.

[104] J. Rowland et al. Status of the diamond fast orbit feedback system. In *Proceedings of ICALEP07*, 2007.

[105] J. Safranek. Experimental determination of storage ring optics using orbit response measurements. *Nucl Inst. and Meth. A*, 388:27–36, 1997.

Bibliography

[106] T. Schilcher. RF applications in digital signal processing. In *Proceedings of the CERN Accelerator School on Digital Signal Processing*, 2007.

[107] H. Schmickler et al. A multi-TeV Linear Collider based on CLIC Technology. Technical Report CLIC-2012-007, CERN, 2012.

[108] D. Schulte. *Study of Electromagnetic and Hadronic Background in the Interaction Region of the TESLA Collider*. PhD thesis, Universität Hamburg, 1996.

[109] D. Schulte. Emittance Preservation in the Main Linac of CLIC. In *Proceedings of the 1998 European Particle Accelerator Conference (EPAC98)*, pages 478–480, 1998.

[110] D. Schulte. Pulse Shaping and Beam-Loading Compensation with the Delay Loop. Technical Report CLIC-Note-434, CERN, 2000.

[111] D. Schulte. Main Linac Beam Dynamics and Specifications. In *CLIC08 Workshop*, 2008. http://indico.cern.ch/conferenceDisplay.py?confId=30383.

[112] D. Schulte. Main linac. In *Proceedings of the ILC School 2009*, Beijing (China), 2009. www.linearcollider.org/school/2009/.

[113] D. Schulte et al. The Tracking code PLACET. Technical report, CERN. https://savannah.cern.ch/projects/placet.

[114] D. Schulte and R. Tomás. Dynamic effects in the new CLIC main linac. In *Proceedings of the 2009 Particle Accelerator Conference (PAC09)*, 2009.

[115] A. Sery. Ground motion model in LIAR. www.slac.stanford.edu/~seryi/gm/model/. SLAC.

[116] A. Sery. Stability and Ground Motion Challenges in Linear Colliders. In *the 33th Advanced Beam Dynamics Workshop (NANOBEAM02)*, 2002.

[117] A. Sery, L. Hendrickson, and G. White. Issues of stability and ground motion in ILC. In *the 36th Advanced Beam Dynamics Workshop (NANOBEAM 2005)*, 2005.

[118] A. Sery and O. Napoly. Influence of ground motion on the time evolution of beams in linear colliders. *Phys. Rev. E*, 53:5323, 1996.

[119] A. Sery and T. O. Raubenheimer. Ground Motion Model of the SLAC Site. In *Proceedings of the 22th International Linac Conference (Linac 2000), SLAC-PUB-8595*, 2000.

[120] V. Shiltsev. Observations of random walk of the ground in space and time. *Phys. Rev. Lett.*, 104(23):238501, Jun 2010.

[121] V. Shiltsev. Review of observations of ground diffusion in space and in time and fractal model of ground motion. *Phys. Rev. ST Accel. Beams*, 13(9):094801, Sep 2010.

Bibliography

[122] S. Simrock. Control theory. In *Proceedings of the CERN Accelerator School on Digital Signal Processing*, 2007.

[123] S. Skogestad and K. Havre. The use of RGA and condition number as robustness measures. *Computers & Chemical Engineering*, 20:1005 – 1010, 1996.

[124] S. Skogestad and I. Postlethwaite. *Multivariable Feedback Control: Analysis and Design*. Wiley-Interscience, 2005. ISBN: 0-470-01168-8.

[125] J. Snuverink et al. Status of Ground Motion Mitigation Techniques for CLIC. In *Proceedings of the 2nd International Particle Accelerator Conference (IPAC11)*, 2011.

[126] St. Stapnes and D. Schulte. CLIC status and outlook. *Proceedings of Science*, 2012. to be published.

[127] G. Stein. Respect the Unstabe. *IEEE Control Systems Magazine*, 23(2):12 – 25, 2003.

[128] R. J. Steinhagen. Feedbacks on tune and chromaticity. In *Proceedings of the 8th European Workshop on Beam Diagnostics and Instrumentation for Particle Accelerators (DIPAC07)*.

[129] R. J. Steinhagen. *LHC Beam Stability and Feedback Control*. PhD thesis, Technische Hochschule Aachen, 2007.

[130] R. J. Steinhagen, S. Redaelli, and J. Wenninger. Analysis of Ground Motion at SPS and LEP - Implications for the LHC. Technical Report CERN-AB-2005-087, European Organization for Nuclear Research (CERN), 2005.

[131] I. Syratchev. Efficient RF power extraction from the CLIC power extraction and transfer structure (PETS). Technical Report CLIC-Note-571, CERN, 2003.

[132] R. Tomás et al. Summery of the BDS and MDI CLIC08 working group. Technical Report CLIC-Note-776, CERN, 2009.

[133] T. Touzé. *Proposition d'une méthode d'alignement de l'accélérateur linéaire CLIC*. PhD thesis, Université de Paris-Est, 2011.

[134] G. White, N. Walker, and D. Schulte. Design and simulation of the ILC intra-train orbit and luminosity feedback systems. Technical report, EUROTeV, 2006. EUROTeV-Report-2006-088.

[135] H. Wiedemann. *Particle accelerator physics*. Springer, 2007. ISBN: 3-54-049043-4.

[136] K. Wille. *The Physics of Particle Accelerators an introduction*. Oxford University Press, 2000. ISBN: 0-19-850549-3.

[137] L.H. Yu. Real Time Closed Orbit Correction System. In *Proceedings of the 1989 Particle Accelerator Conference (PAC89)*, 1989.

[138] F. Zimmermann. Tutorial on Linear Colliders. Technical Report CERN-SL-2000-069, CERN, 2000.

i want morebooks!

Buy your books fast and straightforward online - at one of world's fastest growing online book stores! Environmentally sound due to Print-on-Demand technologies.

Buy your books online at
www.get-morebooks.com

Kaufen Sie Ihre Bücher schnell und unkompliziert online – auf einer der am schnellsten wachsenden Buchhandelsplattformen weltweit! Dank Print-On-Demand umwelt- und ressourcenschonend produziert.

Bücher schneller online kaufen
www.morebooks.de

VDM Verlagsservicegesellschaft mbH
Heinrich-Böcking-Str. 6-8   Telefon: +49 681 3720 174   info@vdm-vsg.de
D - 66121 Saarbrücken   Telefax: +49 681 3720 1749   www.vdm-vsg.de

Printed by Books on Demand GmbH, Norderstedt / Germany